MECHANICAL MIXING MACHINERY

By LEONARD CARPENTER, B.Sc., A.I.C.

LONDON: ERNEST BENN LTD.

8 BOUVERIE STREET, E.C. 4

1925

NEW YORK
D. VAN NOSTRAND COMPANY
EIGHT WARREN STREET

Richard Clay & Sons, Ltd., Printers, Bungay, Suffolk

PREFACE

THE subject of this book, hitherto, has not been dealt with separately, although reference to mixing and descriptions of various types of mixers and mixing machinery will be found in all standard works on Chemical Engineering. The references, however, are usually of a somewhat brief character, as indeed they must be owing to exigencies of space. Moreover, mixing is a process, which, up to the present, has only been treated in an empirical way; unlike such processes as crushing, grinding, separation, which admit of some degree of mathematical treatment as regards size of plant for a given output, strength of materials, and power of operation. This is not to say that a careful study of conditions in designing a mixer for a given purpose will not be amply repaid. But such a study must, of necessity, be mainly of an experimental nature on the principle of trial and error.

This book does not claim to be an exhaustive work on the subject. It is addressed more especially to the student of chemical engineering, to the young works chemist anxious to study the art of transferring a laboratory process to the works. In the brief space at disposal it has only been possible to deal with the problems of mixing in comparatively few industries, and in these only in a somewhat superficial manner. These industries are not all, strictly speaking, chemical; yet chemists are, or may be, advantageously employed in all of them. This might indeed be said of every industry known. But it will be sufficient to say that the industries referred to involve typical mixing problems, and the machines described may furnish ideas for handling problems in other industries.

The author wishes to thank the various makers of mixing machinery who have kindly furnished him with assistance in the way of literature, information, and blocks for reproducing illustrations, and also Mr. N. A. Lloyd, and Mr. J. W. Fagan of the Derbyshire Silica Firebrick Co., Ltd., and Mr. Walter Emery of the Stoke-on-Trent School of Science, for their assistance and advice in the preparation of the manuscript.

L. C.

BUXTON.
February, 1925.

CONTENTS

LIST OF ILLUSTRATIONS

1*

MECHANICAL MIXING MACHINERY

CHAPTER I

INTRODUCTION

The Principles of Mixing.—Operations involving mixing in the chemical and allied industries, have for their object the bringing of two or more substances into more or less intimate contact, either for the purpose of promoting chemical action or of simply providing a homogeneous mixture. In either case, the more intimate the contact the more satisfactory the product. A homogeneous mixture may be defined as one any portion of which is a fair sample of the whole; so that from wherever a sample may be taken in the mass it will always contain the same constituents in the same proportions. As may be imagined, such a perfect mixture can never be realised in practice, but with certain types of machines it can be closely approached, more especially in systems containing a liquid phase. In the case of dry mixings, a perfectly homogeneous system, even if it could be obtained, would be unstable, as any movement of the mass would set up a sifting action, the particles of the finer material tending to pass through the interstices of the coarser and thus to separate from them.

11

This tendency is very much less in a wet mix, owing to the viscosity of the liquid hindering movement, and the capillary action tending to keep the particles of solid in contact with one another.

Taking now the hypothetical case of a perfectly homogeneous mixture composed of two constituents A and B in equal proportions by weight; then, if A and B are both of the same density, each particle of A must lie alongside a particle of B. Similarly, if the proportion be two of A to one of B, then two particles of A will lie alongside each particle of B throughout the mass. These statements are made on the assumption that the particles of both A and B are all of the same size. Suppose now the relative densities of A and B are different, and that A is twice that of B. Still assuming that the particles are all of the same size, then, in this case, there will be two particles of B to each one of A associated together. Where the relative densities do not bear a simple numerical relation the case becomes more complicated; for example, if A has a density of 1·2 and B a density of 1·0, then 6 particles of B will be associated with 5 of A.

It will thus be seen that such perfect association of particles can only be approached by the most intensive action in a mixer. In fact it is probably only in the type of machine known as the colloid mill, to which further attention will be directed, that such intimate incorporation can be obtained.

The principles favouring a close approach to perfect mixing may be enunciated as follows :—

(1) The particles of each constituent must be as small as possible. By this means the association of the correct number of particles together may be the more nearly approached.

(2) The particles of each must be of approximately equal size, otherwise perfect homogeneity cannot be approached.

(This condition may not be practicable in every case. For other reasons, moreover, it may not be desirable; for example, in the manufacture of refractories, the mix as prepared in the mill contains particles which may vary from $\frac{1}{4}$ inch down to colloidal dimensions. This is necessary in order to provide a body combining toughness with porosity. Here, other necessary properties of the mix militate against perfect association of the particles.)

To these two principles may be added a third upon which depends more the stability of the mix when made, rather than the actual attainment of homogeneity, although the latter is affected by it. Namely, the densities of A and B must be approximately the same, or at any rate of the same order. The effect of different densities has already been considered when dealing with the association of particles. In addition, the tendency of the constituents to separate by sifting action is greater where the densities show great disparity. If the sizes of the particles differ considerably, and the smaller particles are of greater density, the tendency to separation will be enhanced, but if the reverse, it will be diminished. In a large number of cases the density of a body is greater the greater the hardness, owing to a more compact texture. Consequently, when two materials are ground together for the purpose of mixing, if one is of greater hardness than the other, the softer material will form the bulk of the finer and less dense material, and such a mixture will show less tendency to separate than one in which the finer material is the more dense.

In such cases it is generally better, if very intimate mixing is desired, to grind each constituent separately to approximately the same size, and incorporate them in a machine whose sole function is mixing.

Types of Mixtures.—Mixtures may be divided into two main classes, viz. homogeneous and heterogeneous. In order to avoid confusion in the term " homogeneous " as used in the previous paragraph, and now, it may be explained that " homogeneous " is now applied to mixtures each constituent of which is in a similar phase, for example, all solids, or all liquids mutually soluble. Heterogeneous mixtures contain constituents in different phases, for example, solid and liquid, liquids mutually insoluble. In such systems, one or more phases will be disperse and one continuous; it is possible to pass through a continuous phase without passing through any of the others, whilst the particles of the disperse phase are all separated from each other by the continuous phase, and it is impossible to pass through the mass without entering the latter.

Considering now the applications of the afore-mentioned principles to homogeneous and heterogeneous mixtures, it will be readily seen that conditions (1) and (2) are more readily satisfied in the case of the former than in the case of the latter. Taking the case of a thick suspension or cream such as milk of lime, the solid, disperse phase consists of particles of hydrate of lime of measurable, if microscopic, dimensions, whilst the water, the continuous phase, consists of particles of sub-microscopic, molecular dimensions. The nearest approach we can get to a perfect mixture is that in which the particles of lime are evenly distributed throughout the water, so that from whatever part of the mass we draw a sample it will always contain the same percentage of lime and water. As prepared in the ordinary way, such a mixture will tend to separate, the lime gradually sinking to the bottom, and in order to maintain a constant mixture some stirring device is necessary.

The conditions (1) and (2) are nearly perfectly satisfied in the case of miscible liquids, because here all particles are of molecular dimensions. Moreover, the molecular movement going on tends to prevent separation, which does not take place even in the case of liquids of such widely differing densities as ether and chloroform. Mixtures of mutually soluble liquids are absolutely permanent. Such permanency can be obtained in heterogeneous systems if the disperse phases be reduced in a colloid mill to particles which begin to approach molecular dimensions. In the ordinary way, if we shake up petrol and water together we obtain a temporary emulsion, but this quickly separates as the particles of each phase coalesce. If, however, the mixture be disintegrated in a colloid mill, the particles are broken up into such minute dimensions that they do not coalesce, especially if some deflocculant be added which, by coating the particles, lowers the surface tension and prevents their reuniting.

CHAPTER II

OPERATIONS OF MIXING

BEFORE describing the types of machinery suitable for mixing, it will be well to consider first the procedure of mixing by hand. The usual plan to be adopted consists in spreading each constituent out in an even layer one on top of another; the proportion by volume of each constituent will then be in proportion to the depth of each layer. The operator then proceeds to turn the mass over with a shovel, thus bringing the bottom layer to the top, and *vice versa ;* in so doing a partial mixing is effected, which becomes more intimate the more the mass is turned over. The operation is continued until the whole mass appears homogeneous; if the constituents are of different colours or shades, it becomes much easier to judge when this condition is reached, as any concentration of one particular constituent in any part of the mass is evident by an appearance of streakiness or patchiness. Hand mixing is still employed for small quantities, especially when one or more of the constituents is of a light and dusty character, as less dust is likely to rise than in a mechanical mixer unless this is covered in and some arrangement made for the collecting of the dust. A familiar example is the operation of mixing small quantities of mortar. It is also employed in the fertiliser industry for blending small orders for compound manures of non-standard composition, when the quantities are not sufficiently large to justify cleaning out a mechanical mixer for the purpose.

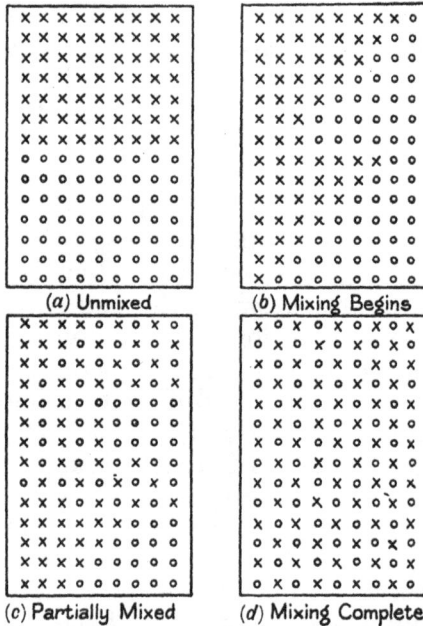

(a) Unmixed

(b) Mixing Begins

(c) Partially Mixed

(d) Mixing Complete

FIG. 1.—MIXING IN FOUR STAGES.
Material A thus : ×××
Material B thus : ○○○

Mechanical mixers are designed to carry out the above operation more rapidly and with less expenditure of labour. Fig. 1 shows a mix in various stages of preparation ; in (a) the two constituents A and B are shown, the one on top of the other before mixing has begun. In order to effect mixing, the moving member is designed to draw the particles of B up through the mass of A, whilst the latter is at the same time caused to pass downwards, thus producing different stages through (*b*) to (*d*). This may be effected by means of a shaft fixed vertically in the container and carrying blades arranged helically in such a manner as to effect the required action. Or the shaft may be set horizontally so that the blades lift and turn the mass over in the same way as the spade in the hand-mixing operation. The vertical shaft type requires less power to operate, but takes longer than the horizontal shaft type to effect mixing. The proper type to employ can only be decided by the particular conditions involved, such as density and size of the particles, and general texture of the materials involved. In general, it may be said that for solids the horizontal type is usually preferred, whilst for the liquids the vertical type is preferable. We may now proceed to discuss in detail various operations of mixing and to analyse the peculiar conditions involved in each.

Two or more Solids.—In order to incorporate thoroughly two or more solids it is usually necessary to reduce each to a more or less fine state of division; this is in accordance with the first principle laid down in Chapter I. In certain cases, this may not be desirable for other reasons, as in the case of a clay mix in the manufacture of firebricks, where a certain proportion of coarse calcined materials, or "grog," is added to the raw clay to give strength and porosity, and to reduce shrinkage. The grog plays a similar part to the coarser particles in a concrete aggregate, and in consequence must not be too fine. A considerable proportion of fine material is necessary in the raw clay, however. Such a mixture cannot strictly be called homogeneous, because it would be possible to pick out from the mass a piece consisting almost entirely of grog, or without any grog at all, which could not be said to be representative of the whole. Taking, however, the case where a high degree of homogeneity is desired, the first process will consist in reducing each constituent to a certain degree of fineness, and in accordance with the second principle laid down in Chapter I this must be approximately the same for each. Where the constituents are of a similar degree of hardness, and consequently become crushed to the same degree of fineness when subjected in a mill for the same periods of time, it is possible to combine the operations of grinding and mixing; a variety of machines are made which combine both functions. In fact, although the principles of grinding and mixing are entirely

different, it is impossible to deal with the
latter without introducing the former in
some shape or form.

Where the materials are of a different
order of hardness it becomes essential to
grind separately and incorporate in a
machine whose sole function is mixing.   An
example of this may be described in the
mixing of basic slag and mineral phosphates
in the preparation of the so-called slag
phosphate, or " enriched slag."   Of these,
the slag is by far the harder of the two
materials.   If an attempt be made to grind
and mix the two simultaneously in a ball
and tube mill installation, the phosphate is
quickly reduced to powder and will pass out
of the discharge trunnion of the tube mill
before the slag.   If the discharge be closed
up to prevent this, the mill has to be run a
much longer period than is necessary to
grind the phosphate before the slag is
reduced to fine powder required in a fertiliser of this description.   Moreover, the mill
does not function so efficiently, owing to the
cushioning action of the ground phosphate,
the coarser particles of slag tending to
become embedded therein under the action
of the balls.[1]

Whether the operations of grinding and
mixing are carried out separately, or together in the same machine, the latter

[1] See Fig. 4, p. 38.

consists in turning over the mass, causing one constituent to constantly replace another until conditions of homogeneity are sufficiently approached for the purpose involved. Such a mixing will be fairly stable and permanent if conditions (1) and (2) (Chapter I) have been satisfied to a sufficient degree. Any considerable difference in grain size, especially if associated with a greater density on the part of the finer particles, will tend to produce " sifting action " as described, and consequent separation.

Two or more Miscible Liquids.—When one liquid is poured upon another with which it is mutually miscible, in a vessel, the denser will sink to the bottom and form a separate layer. The molecular movement of each, however, will slowly cause it to diffuse into the other, until in time a homogeneous mixture will result. Such a process would, however, be extremely slow, and may be hastened by heating, the convection currents quickly effecting mixing. Or liquids may be agitated, either mechanically by means of a paddle, or by forcing compressed air or steam through the mass.

Conditions (1) and (2) are probably more fully satisfied in this case than in any other without any previous preparation, and consequently such mixings do not present any great difficulty. Examples include the blending of oils by compressed air, the mix-

ing of various organic compounds with
strong sulphuric acid in processes of sulphonation; numerous other examples will
occur to the reader.

A very simple type of agitator for such
mixers consists of a hollow truncated cone
or cylinder mounted concentrically with the
shaft, to which it is joined by blades or webs
arranged helically. The effect of these is
to draw the liquid up from below through
the cylinder or cone, and pass it towards
the surface, whence it returns by gravity on
the outside of the rotating member. If this
member be in the form of a cone it is usual
to arrange it with the smaller end lower.

The circulating action of an agitator of
this description renders it unnecessary to
make use of the arms on the rotor to sweep
the full diameter of the vessel. Such an
arrangement might effect more rapid mixing, but at the expense of more power.
Moreover, the deep vortex which would be
produced might cause the liquid to overflow
the sides of the vessel if this were open,
unless it were not more than half filled.

An extremely simple method of mixing
two mutually soluble liquids consists in
running them from separate tanks into a
common chute which discharges into a
third or mixing tank. This is a very convenient method of mixing solutions of substances which react to form a precipitate.
An example is the preparation of lead

arsenate by mixing solutions of lead nitrate or acetate and sodium arsenate, but may be applied to almost any similar case. A familiar and homely example well known to housewives is the preparation of French coffee by pouring coffee and milk into the cup simultaneously. It is probably the extremely intimate mixture of the coffee and milk produced in this manner that accounts for the excellent flavour.

Solid and Liquid.—Mixtures of solids and liquids in which the former are insoluble in the latter belong to the class of heterogeneous systems. They may be subdivided into two : (a) pastes, when the solid preponderates; and (b) creams, slurries, or slips, when the liquid preponderates.

With (a), in order to obtain a proper degree of homogeneity, it is necessary, in accordance with condition (1), to have the solid in a fairly fine state of division, or at any rate a sufficient part of it to hold the coarser particles up; as in the case of a clay paste used in the ceramic industries, where the mixture may be composed of, say, 40—50% of fine material with comparatively coarse particles of calcined clay or flint. Provided the liquid phase is not volatile, a paste is quite stable and retains its original degree of homogeneity indefinitely, the liquid being retained in the interstices of the solid by capillary action.

The preparation of such a paste is by no

means so simple as might at first be imagined. In the clay industries the solid constituents are frequently first mixed dry, then sprinkled with water and left to soak for varying periods. This process of " ageing " or " souring," as it is called, is supposed to be, to a certain extent, a bacterial action, but there is no doubt that the gradual soaking of the water through the mass under the action of gravity plays an important part, and saves considerable power when the stiff paste is forced through the die of the pug-mill, besides rendering it more homogeneous. In the case of extremely short, siliceous clays, unless this " ageing " process has been carried out properly, there is risk of breaking the knives of the pug-mill.

In the case of (b), where the liquid phase predominates, the problem lies not so much in obtaining a proper mixture as in preventing separation when once obtained. If the solid be of greater density than the liquid, it will tend to fall to the bottom, and if of less density, to rise to the top. Only in the extremely rare case of the solid being exactly the same density as the liquid can the system be said to be stable. If, however, the particles of the solid be of such dimensions that the ratio of their area to their mass is very great, i.e., if the particles be extremely small, separation may be hindered to such an extent as to render the suspension practically permanent. Here we have a case where condition (1) must be well satisfied. Such permanent creams and suspensions can only be produced by means of special types of machine such as the colloid mill, since the particles must be of more or less colloidal dimensions.

In the pottery and cement industries, where slips or slurries are used, stability is maintained by means of slow moving stirrers in the stock tanks for the mixtures.

Two or more Immiscible Liquids.—The same conditions apply in this case as to stable suspensions. The disperse phase or phases must be comminuted to such small dimensions that the ratio of area to mass is of such an order as to make the resistance to movement in the continuous phase as great as possible. Here again the colloid mill is the only means whereby such conditions may be realised mechanically. It is possible, however, to prepare stable emulsions by the addition of emulsifying agents. The emulsification of creosote oils by means of soap in the manufacture of disinfectants is an example of this.

It should be mentioned that it is not merely the relation of area to mass which renders dispersions of the above character stable, but also the Brownian movement exhibited by all colloidal particles, due to the electrical charges thereon.

CHAPTER III

PERHAPS the great majority of mixing problems involve the mixing of solids. Such processes may be divided into three : dry, semi-dry, and wet. The addition of water or other liquid, where this does not affect the materials or the product in any detrimental way, is frequently an advantage It lubricates the movement of the particles upon one another and thus absorbs less power for the mixing operation. This is more especially the case with dense solids, or those in which the particles tend to interlock and so bind together. It stands to reason that dense bodies will absorb more power than light ones, owing to the greater amount of work required to be done in lifting them.

We may now proceed to describe various types of mixers in use. They may be divided into two main classes : batch mixers and continuous mixers. In the former, the materials are placed in the machine in their required proportions, one after the other, or simultaneously, and after mixing, are discharged by suitable means, and a further batch is put in and mixed, the operation being continued until a sufficient quantity has been made. In the latter, the materials are continuously fed in at one point and withdrawn at another, the mixing gear also acting as a conveyor, passing the materials through the machine. It is possible partially to combine both functions in what may be termed a semi-continuous mixer. The materials are introduced in batches but withdrawn continuously, fresh material being added before the previous batch has been completely withdrawn. Such an arrangement can hardly be recommended from the point of view of accuracy, as there is risk of contamination of the outgoing product with the first constituent of the next batch. For rapid output where a high degree of constancy of composition is not important, however, it has advantages.

The advantages of the continuous mixer from the point of view of rapid output are obvious; in addition, there is no power wasted in revolving the mixing gear idly between the addition of the batches; although this power in general is quite small in itself, it may amount to quite a large item with very large outputs and should not be left entirely out of consideration under such circumstances.

To obtain really good results with continuous mixing, it is important to have an accurate feeding mechanism which will measure the constituents in the right proportions. In some cases, the correct quantity of each constituent is brought and dumped beside the mixer in sufficient quantity to last for some time. The operator, roughly gauging the right proportions, feeds the materials by alternate shovelfuls into the machine until the supply of each is exhausted; if his gauging has been fairly correct, he will exhaust each heap at the same time, but if not he will have a supply of one or more left over; he must then feed the whole of the material through the machine again, adding the surplus material gradually so as to incorporate it with the rest of the batch. The extra time and expense involved in such a procedure are obvious. In practice, however, the operator becomes skilled in gauging the quantities, and any inaccuracies in the gauging are evened out in the mixer if this be of large capacity. One of the advantages of continuous mixing, however, lies in the ability to use a small machine absorbing comparatively little power for large outputs, so this advantage would be greatly discounted when using a larger machine.

In spite of its drawbacks, the above method of feeding continuously by hand, and relying upon the judgment of the workers for accuracy, is still used to a great extent, especially in the fertiliser industry for preparing compound manures. These " Compounds " have to come up to a strict guarantee which is required by law to be stated on the containers, so that it would seem highly inadvisable to make use of such haphazard measures. In practice, however, practically all compounds have to undergo an ageing period in heaps, and before bagging are subjected to a further disintegrating and screening process; this effects further mixing, evens out "pockets," and renders the whole mass reasonably homogeneous in most cases.

To obtain really accurate and even composition in one operation, however, it is essential to use some kind of measuring device. For liquids, this is comparatively simple, as, provided the head remains constant, the rate of flow through an orifice of fixed size is constant. With solids, however, it is much more difficult; in fact, with materials of fibrous or thread-like texture, practically impossible. Fine, dry powders of constant density behave to a certain extent like liquids and can be made to flow through orifices, but such an arrangement can seldom be recommended, as it would require constant attention to avoid "hanging up," and consequent stoppage. The best arrangement is one in which the solid is positively fed, and an example of this is the measuring table illustrated in Fig. 2. The material is fed either by hand or mechanical means into the hopper, A, whence it falls on to the rotating table, B, and is continuously diverted by the knife, C, which cuts it off in a definite manner and causes it to fall off at D. Any desired proportion may be measured, according to the depth to which C is set on the measuring table.

Two or more tables with their respective
hoppers may be arranged, one above
another, rotating on a common shaft, so as
to measure proportions of more than one
constituent, which are caused to fall off at
points one above another into the mixer.

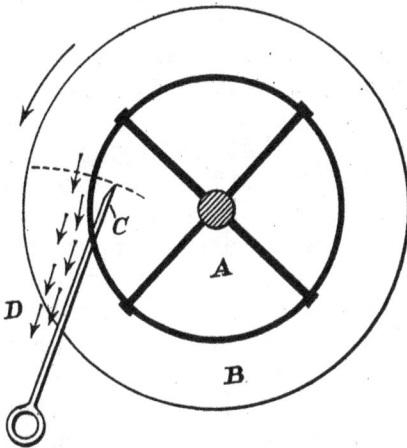

FIG. 2.—TABLE MEASURING MACHINE.

Provided the hoppers are kept filled above
the level of their lower edges, the proportion
measured off will be the same, however
much is fed into them; so that the feed
may be more or less intermittent according
to the size of the hopper.

Other types of measuring machine include
screw conveyors with tapering threads
which can be set at various positions in the
feed trough, and rotating valves. The

latter have the disadvantage that the feed and discharge are by gravity, and thus with certain materials " hanging up " is liable to occur; they appear, however, to give satisfaction in many cases.

The measurers have to be calibrated separately for different materials. In spite of the obvious advantages of continuous mixing, engineers in this country seem to prefer mixing in batches, more especially in regard to concrete. This is probably due to the fact that measuring devices do not as a rule provide the degree of accuracy required for such purposes.

Turning now to actual types of mixers in use, one of the oldest and still very popular machines is the edge-runner or pan mill, of which a diagram in plan is shown in Fig. 3. It consists of a circular pan, A, in which are fixed broad rollers, BB, which are free to rotate independently of one another upon a shaft, C. Either the pan, A, may be rotated, the rollers, BB, turning in opposite directions to one another by friction with the pan bottom; or the pan may be kept stationary and the rollers caused to travel round it, movement being imparted through the spindle, D, by means of the crown and bevel wheels, E and F; in this case, the spindle, D, is keyed solid to the shaft, C. If the pan is to be rotated, D passes through a bearing in C and is keyed to the centre of A; in

this case the pan itself rotates upon, and is
supported by a massive bearing beneath.
This bearing must be of a very substantial
nature, as it has to take the weight of the
pan, of cast iron or steel and usually very
heavy, together with the weight of the
contents; for this reason the stationary

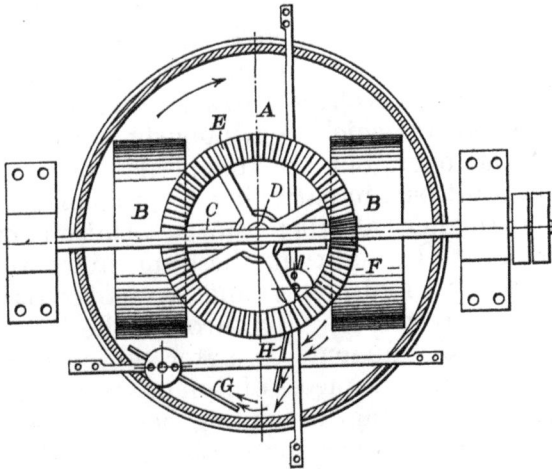

FIG. 3.—LARGE PAN MILL.

pan is sometimes preferred, but it is not so
convenient for feeding, as the rollers are
continually passing the feed point, and
material is liable to be deposited upon them,
and, if of a sticky nature, is liable to adhere.

In order to keep the material in the path
of the rollers, from which it tends to pass
outwards by centrifugal action, scrapers, G
and H, are provided. These scrapers are

2

stationary when the pan rotates, but move round with the rollers when the pan is fixed.

These machines are grinding mills in the first place, but are extremely useful for mixing heavy materials which are not affected by the crushing action of the rolls, such as rocks, clay and sand, and for this reason are largely used in the refractories industry and for mixing mortar.[1]   The mixing action is effected to a great extent by the scrapers which continuously turn the material over, but the rollers also play an important part in flattening it out again and thus further altering the association of particles.   In addition, these machines are practically fool-proof, and cannot jamb through overloading, or through the introduction of fragments of materials too hard to crush, such as pieces of iron or steel.   The rollers are hung in sliding bearings, and simply climb over materials too hard to crush.   If the pan be overloaded, it simply fills up, the rollers revolving upon the surface of the material.

Pan mills can be used as continuous or batch mixers.   In the former case the pan is perforated with holes of rectangular or trapezoidal form, through which the material falls as it becomes sufficiently ground, and drops to the boot of an elevator,

[1] Mortar mills are frequently only provided with one roller.   In this case the pan rotates.

whence it may be raised to a screening
plant, provision being made for returning
the tailings, *via* a chute, to the pan.   When
used as a batch mixer, the pan has a solid
bottom and the material is ground for a
sufficient period to effect the desired mixing.
It may then be discharged whilst still
running by inserting an open-ended trough
or chute, up which the material is forced as
the pan travels round.   In some cases the
pan has a rotating bottom only, the sides
being stationary; a door is provided in
the latter, which can be opened when
the material is discharged by altering the
position of the scrapers so as to sweep the
material out.

With a fixed pan, mechanical discharge
may be arranged in a similar way by means
of a door in the side or bottom of the pan;
the position of the scrapers is altered so
that the material is swept towards the
opening and discharged.

Another type of pan mill is provided with
a pan having a conical bottom which slopes
slightly downwards all round towards the
middle.   Each roller is formed as the
frustum of a cone, the larger diameter being
innermost.   It is claimed that the grinding
and mixing actions are improved thereby
and the tendency towards clogging is diminished.[1]   In some cases the rolls are sup-

[1] Patent Specification 191208, Grinding, Crushing
and Mixing Mills, J. Wass.

ported a few inches above the pan so that a certain depth of material is required before they will revolve. This, however, is more for grinding than mixing purposes and so does not call for treatment here.

Pan mills may be constructed of cast iron or steel. Where contamination with iron is to be avoided, the rolls and the bottom of the pan may be made of granite, and the scrapers of lignum vitæ or other hard wood. For really good results both pan and rollers should be turned perfectly true.

With regard to motive power, this may be by belt from a line shaft or motor, to fast and loose pulleys mounted on the end of the shaft carrying the bevel wheel, F. This shaft may even be driven directly by means of a crank mounted on the end, attached to the connecting rod of a steam engine. In other cases the pan may be under-driven by crown and bevel wheels, but this method is not to be recommended from the point of view of cleanliness and accessibility, except perhaps for small machines mounted above the floor level where the driving pinions are readily accessible for lubrication and adjustment.

Pan mills are made and used in all sizes, from small machines with a capacity of a few hundredweights per hour up to eight tons, and are used in a great variety of industries. Their simplicity and fool-proof character are a strong recommendation.

They may be used for both dry and wet mixing, but in the latter case, if the material be of a pasty consistency, it is desirable to provide scrapers to the rolls to remove any adherent matter.

Other types of grinding mills sometimes used as mixers are ball and tube mills. The principle of these is similar, namely, a rotating cylinder containing a number of balls of steel or flint. The cylinder is lined with some hard material such as cast steel, or quartzite, and grinding is effected by the rubbing and pounding of the balls on the material as the cylinder rotates. The difference between the ball and the tube mill lies in the arrangement for discharge. The former, which is usually of shorter barrel than the latter, contains a series of stepped plates, as shown in Fig. 4. Revolving in the direction of the arrow, the balls constantly roll down and fall from plate to plate, the materials thus being mixed and ground simultaneously. The discharge is *via* holes drilled in the plates and the material passes over screens surrounding the whole cylinder. The tube mill is usually of longer barrel, and is simply a hollow cylinder; it is fed either through a manhole when stationary, if used as a batch machine, or through a hollow trunnion if used continuously, the discharge being effected in a similar manner.

These mills may be used for both dry

and wet mixing, but if the latter, the liquid must be present in sufficient quantity to produce a slurry; if not, a thick paste is formed which coats the balls and clogs the action to such an extent as to render it

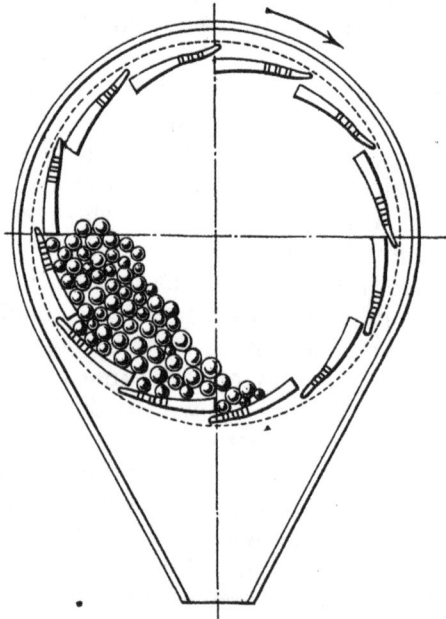

Fig. 4.—Ball Mill.

useless. Where mixing only is required, the balls may be replaced by projecting arms, which turn the material over.

As mixers, however, ball and tube mills have a far more limited application than pan mills. In the manufacture of cement, however, by the dry process, the tube mill is used for the purpose of grinding and mixing. (See Chapter V.)

Crushing rolls, consisting of a pair of rolls, either plain, grooved or fluted, and revolved in opposite directions in close proximity, have certain limited uses as mixers, notably in the rubber industry, and to a certain extent in the clay industry. Their action is to squeeze the materials into contact, but when used alone for this purpose, repeated passage of the materials between them is necessary, or else they are used in conjunction with other machinery. (See Mixing in the Ceramic Industry, Chapter VI.)

As previously stated, grinding and mixing can only be carried out simultaneously with success when the materials are of a similar degree of hardness.

Turning now to machines whose sole function is mixing, we find that there is an enormous variety of these. One of the commonest and most popular is the horizontal shaft mixer which is suitable for dry mixing, or wet mixing with a preponderance of solid (*i.e.*, pastes). It consists of a trough of semicircular section containing a rotatable, horizontal shaft carrying a complete open worm, or else helical blades, the latter being arranged like an interrupted worm. A continuous solid worm would be unsuitable, as the machine would then act simply as a conveyor which would push the material to one end, where it would simply bank up and overflow if the trough were open, or wedge up and stop the machine if closed. The open worm simply conveys the material until it rises above the level of the worm blade, when it falls back again, the action being continually repeated and mixing effected. The interrupted worm type works in a similar manner. The action consists entirely in picking up and turning over the material. Where the mass is liable to be rotated as a whole, as in the case of sticky or glutinous materials, baffles are arranged between the blades, or, what amounts to the same thing, the blades revolve in pockets. Not only is it important to prevent rotation of the batch as a whole, but also to reach the whole of it, and for this reason the clearance between the tips of the blades and the inside of the trough should be as small as possible without risk of jambing.

Horizontal mixers are sometimes constructed with two shafts set parallel to one another and revolving at different speeds in opposite directions. The ratio of speeds may be as four to five. This arrangement effectively prevents rotation of the batch as a whole and effects more thorough mixing.

A typical example of such a mixer is Fawcett's double-shafted clay-mixer, made by Thomas C. Fawcett, of Leeds. It is supplied either with or without an automatic feed and patent water control, and may be used as a batch or continuous machine, but is primarily intended for the latter. It is largely employed in the brick-making industry, and the output, which is entirely automatic, may be adjusted to the capacity of the brick press.

Horizontal shaft mixers may be provided with jackets for the circulation of steam, hot or cold water, or oil, so that the batch may be heated or cooled during mixing; in addition, the shaft may be made hollow for the same purpose. An example of this is the " Atomixer," made by the Keenok Co., Ltd. This machine consists of a barrel carrying a hinged lid which may be turned back for cleaning and inspection; mixing is, of course, carried out with the lid closed. The single hollow spindle carries double-headed beaters, which, when the shaft is revolved at speeds varying from 150 to 600 revolutions per minute, subject the material to an intensive kneading action. In addition to this, a backward and forward oscillating motion is imparted to the mass, the latter being always a little greater than the former, so that the material is slowly carried forward, and the process thereby rendered continuous. This action is effected by the special manner in which the beaters are machined, and this can be

2*

varied according to the class of material to be dealt with.   Although primarily a continuous machine, it may be used as a batch mixer.   It has found considerable application in the chocolate trade, and for the mixing of floor polish, and similar preparations requiring emulsification.   Its action is peculiarly intensive, owing to the high speed of revolution; in fact, the machine stands in an intermediate position between ordinary, low-speed mixers, and the high-speed, disintegrating mills producing particles of colloidal dimensions and running at speeds up to 3000 revolutions per minute (see Chapter IV).   The fifty-six cubic feet emulsifier runs at 300 to 500 revolutions per minute, and requires from three to eight horse-power, according to the consistency of material.

Vertical shaft mixers are used to a very large extent, and take a great variety of forms.   One of the commonest consists of a circular vessel with parallel, vertical sides and a hemispherical bottom; the vertical shaft is concentric with the vessel and carries blades or arms which may be quite short, or may reach almost to the extreme width of it, according to the purpose for which the machine is required.   These arms are usually "staggered" or arranged spirally on the shaft and may be either single, *i.e.*, each arm extending on one side of the shaft only, or double, *i.e.*, the

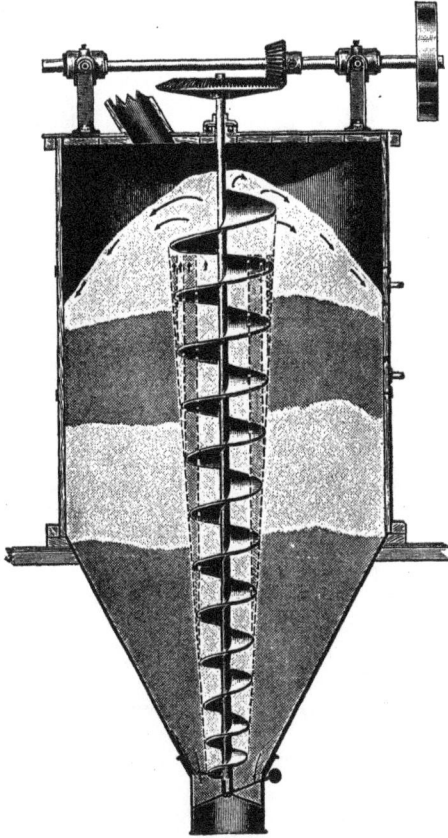

FIG. 5.—GEYSER MIXER.

shaft passing through the middle of the arm. The arms generally take the form of blades which are twisted so as to lie in an oblique plane. The object of this is to give a vertical as well as rotary movement to the materials (see Fig. 15, p. 104).

The blades are sometimes arranged as an open worm or spiral. This lifts the material to a point, from which it then falls back within the ring of the spiral and thus effects the required continual turning-over movement. These mixers are generally employed with semi-solid, pasty materials, creams and suspensions, and for mixing liquids. They are not, however, very efficient, since the action is mainly a rotary one, with the result that the mass is to a great extent merely revolved so that the layers of the different constituents remain more or less separate. To effect thorough mixing requires a vertical motion as well, and to this end numerous types have been designed. A well-known and excellent example of this is the Geyser mixer made by Alfred R. Tattersall & Co., and illustrated in Fig. 5. As will be seen, the vertical shaft carries a solid, tapering worm; this acts as a conveyor, and lifts the material from each layer in proportion to the depth of the layer. In the cylindrical portion of the container the depth of each constituent will be in proportion to its percentage by volume, and consequently as

the worm rotates it withdraws each constituent proportionately and delivers it at
the top. The material withdrawn is replaced by gravity. Thus, if we put a
material A in first, and add a material B,
and rotate the worm, A and B will be withdrawn from their respective layers in proportion to their volume percentages in the
mass, and delivered in a more or less
mixed condition on to the top, until the
supplies of A and B are exhausted and
replaced by the mixture AB; further action
intensifies and finally completes the mixing.

Owing to the worm being tapered, the
largest diameter being at the top and the
smallest at the bottom, each turn of the
spiral lifts more material than the one
below it, with the result that the mass is
kept loose and open, so that the worm
revolves freely, and with far less absorption
of power than would be the case were the
spirals of the same diameter. In fact, the
worm can be said to carry up the material
on its upper surface rather than to revolve
in it. To overcome the friction of the
material absorbs a considerable amount
of power which is only wasted, and this
ingenious but simple arrangement considerably reduces it. This mixer is of the
batch type, and is fed at the top and discharged at the bottom. It is made in a
number of sizes, varying from 2 feet 8 inches
diameter by 9 feet 10 inches in height, to

FIG. 6.—TORRANCE PATENT MIXER.

3 feet 3 inches diameter and 19 feet 6 inches in height, and with capacities ranging from 5 cwt. to 20 tons.

It is essentially a dry mixer and can be used for mixing dry powders of all descriptions, and granular materials such as poultry food and other feeding stuffs, flour and grain.

As in the horizontal mixer, it is possible to use two mixing shafts to intensify the mixing and prevent rotation of the batch. The shafts may be set eccentrically, side by side in a circular container as in the Torrance patent mixer shown in Fig. 6. It will be noted that each agitator is composed of four steel blades which are twisted so that their sides are at right angles to their line of movement; this causes greater resistance than if they were in the line of movement, and consequently the agitation is increased and rendered more effective, and rapid mixing brought about. The housing carrying the agitators can be lifted out to enable a full pan to be replaced by another, consequently delays caused through having to clean pans are largely eliminated. The pan also rotates, and by means of a simple mechanical device can be stopped whilst the agitators are still kept working and the mix run out, thus maintaining mixing action to the last and assuring uniformity of product.

Another Torrance machine is known as the

" perfect " mixer, and is similar to the above
except that only one rotating agitator set
eccentrically is provided. This rotates
in the opposite direction to the pan, thus
doubling the relative speed which would
prevail if the pan were stationary. The
object of setting the agitator eccentrically
is to return the material thrown out
centrifugally, to the centre.

These machines are used to a great extent
in the paint industry for mixing dry colours
in oil, and thinning down paints, but may be
used for practically all mixings of a similar
nature. The ": patent " mixer with two
agitators has a capacity of 30 gallons; the
pan rotates at about 27 revolutions per
minute, and the agitators at 54. The
" perfect " mixer, with single agitator, is
smaller, having a capacity of 12 gallons;
the pan rotates at about 45 revolutions per
minute, and the agitator at 75.

A machine eminently suitable for mixing
liquids with liquids or solids, is the Vortex
mixer made by S. H. Johnson & Co., Ltd., of
Stratford. This machine, an illustration of
which is shown in Fig. 7, is designed to
produce a powerful " vortex " in the fluid.
As previously pointed out, the ordinary type
of mixer with plain agitating arms is somewhat slow and inefficient in action owing to
the movement being mainly a rotatory one,
with the result that the constituents tend
to remain in layers. The shaft, A, rotates

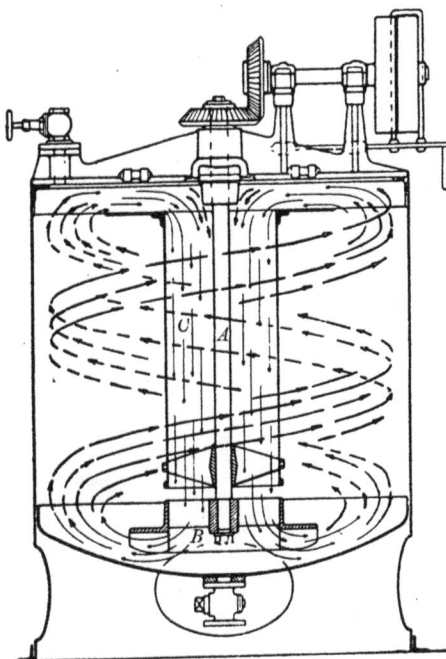

FIG. 7.—VORTEX MIXER.

concentrically in the circular container and carries at the bottom a centrifugal vane, B. As it rotates it throws the fluid out tangentially, imparting a rotary movement, thus causing it to take the spiral path shown by the dotted lines outside the tube, C. Curved plates fitted at the top of the mixer catch the upper layer of rotating liquid and deflect it, and the combined streams descend through the central tube to feed the centrifugal vane at the bottom. The principle is similar to that of the Geyser mixer in that the different proportions of each constituent are drawn from each separate layer, the only difference being that the " vortex " mixer is designed to work on fluids and the " geyser " on dry solids. The action is so powerful that large leaden shot placed in the machine with sufficient water to cover the central tube are carried to the top and circulated with the water. The machine is, therefore, eminently suited for the extraction of such heavy materials as metallic ores, and the mixing of heavy pigments such as barytes. It may also be used for treating syrups and oils with decolorising carbon.

Owing to the absence of baffles, the power consumed is relatively small, but will depend for any given size of mixer on the nature of the material to be dealt with. The " vortex " mixing gear can be fitted to any suitable vessel which may be made in cast iron, steel, or acid-resisting iron, aluminium, and wood. Coils may be fitted for heating or cooling, and the vessel may be made air-tight, so that it may be emptied by compressed air like a montejus or an acid egg.

The type of agitator described in Chapter II, p. 23, for mixing liquids mutually soluble works on the principle of the vortex, but, owing to the swirling movement being only slight, it is not so efficient as the " vortex " mixer just described. For mixing liquids in the processes of nitration and sulphonation, however, it is used to a considerable extent.

The discharging of mixers may be carried out in a variety of ways, depending on the size of the machine and the consistency of the material. Small machines are most simply emptied by tipping them upon trunnions. This is most easily arranged with horizontal shafts, as the pan or container can be made to turn about the same centre as the shaft. When mixers are arranged to work on fluids or on solids which flow readily (e.g., loose, dry powders), the simplest method is to discharge by gravity through a cock or trap at the lowest point. This has the advantage that the mixing gear may be kept running whilst discharging, thus securing uniformity of product. With a horizontal shaft and tipping container the movement of the agitator may be employed to assist discharge.

Compressed air may be used to discharge liquids by rendering the container airtight and arranging the discharge pipe to reach the bottom or preferably to dip into a shallow well, so that complete discharge may be effected after the manner of a montejus. (See description of the " vortex " mixer.)

Continuous discharge in continuous mixers is usually effected by means of an open trap in the bottom of the container at the end opposite to the feed end. The agitating gear also acts as a conveyor and forces the material out.

More individual examples of mixers will be given when dealing with mixing in various industries, but the above describes the main types, and serves to show the principles upon which mixing is based.

CHAPTER IV

INTENSIVE mixing, as the name implies, involves the mixing of materials to a degree of intimacy unattainable with the usual types of mixing machinery. It is a process comparatively new to chemical engineering, having only been introduced during the last few years, but already bids fair to revolutionise certain industries. The process is at present only applicable to wet mixing, and involves the comminution of particles, either of solid or liquid, to colloidal dimensions, so that they remain permanently suspended in the continuous liquid phase by virtue of their enormous surface/mass ratio, and the Brownian movement characteristic of colloidal particles.

It is possible to reduce solids to this fine state of division in the dry state, and mix them with the liquid afterwards, but the usual practice is to effect grinding and mixing in the same machine. Owing to the first principle laid down in Chapter I (reducing the particles to as fine a state of division as possible) being practically fully satisfied, such mixes approach the ideal of homogeneity to a much greater extent than any other. The principle upon which such fine reduction is based consists in subjecting the material which, if a solid, has already been reduced in an ordinary grinding mill to a fine state of division, to a powerful shearing stress between two surfaces set very close together and moving at a relatively high speed, or to a high speed beating action. By this means the particles, whether liquid or solid, are torn asunder and reduced to colloidal or semi-colloidal dimensions. This process enables materials to be ground and mixed in a few minutes, which, with the ordinary type of grinding and mixing machinery, would take many hours, or even days in some cases.

In addition to the shearing action between surfaces set very close together, there may be a powerful beating or disintegrating action by means of beaters mounted upon a rotor revolving at high speed. Such a combined beating and shearing action produces an extremely fine disintegrating and mixing action, it being possible thereby to produce 80% of particles having a diameter of less than 0.1μ ($1\mu = 1/1000$ mm.).

The only machine which embodies both these principles, and perhaps the best known, is Plauson's colloid mill, invented by Dr. Hermann Plauson, of Hamburg, and now being made in this country by Messrs. Mather and Platt, Ltd. It consists of a short, cylindrical casing as illustrated in Fig. 8, which is jacketed for heating and cooling. This carries, mounted eccentrically in its lower portion, a shaft, A, carrying eight beaters, B. Interposed between these beaters are fixed members or anvils, as they are termed; two series of these are fitted, one series, C, being fixed to the bottom of the casing and projecting upwards, and the other series, D, being fixed to the baffle plates, E. The clearance between the

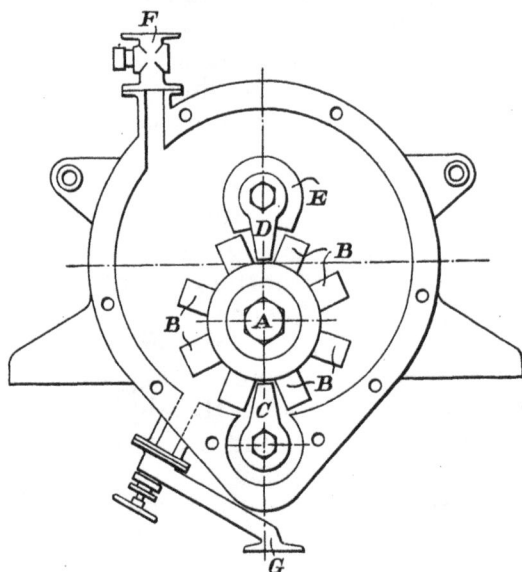

FIG. 8.—PLAUSON'S COLLOID MILL.

beaters and the anvils is 1–2·5 mm. The materials are fed to the machine through the funnel, F. The shaft rotates at 3000 revolutions per minute and upwards, and the tips of the beaters move at 9400 feet per minute and upwards; the number of impacts on the material is approximately

168,000 per minute. It will be readily understood that mixing under such conditions is extremely intimate. The exact *modus operandi* of the grinding action is not fully understood, as the clearance between beaters and anvils is by no means so fine as in mills which work entirely on the film principle.

The baffle plates, E, enclose a portion of the beater track, but do not extend to the lower portion of this, so that ample space is left for free contact with the material. The object of these baffles is to reduce the eddy currents formed which absorb power uselessly; by their means it is claimed that 25% of power is saved.

It will be readily understood that, with the high speeds involved, it is necessary to employ only first-grade materials in construction. The body is made of high-grade, close-grained cast iron, specially treated to withstand corrosion. The beaters are of high-grade steel, securely keyed on to the shaft, which is of forged steel. The rotor is accurately balanced to eliminate vibration at high speeds; it is carried on two sets of roller bearings mounted outside the mill with the driving pulley between. In order to prevent escape of material at the point where the shaft enters the casing, centrifugal throwing discs are provided in conjunction with a gland and stuffing-box. The shaft does not penetrate the other end of the casing, which is bolted on, but may be readily withdrawn without lifting off by simply sliding upon a pair of rails, thus giving access to the interior of the mill for inspection and cleaning. A glass gauge is also fixed to the end plate in order to observe the height of fluid in the mill.

The machine may be used as a batch or continuous mixer. The output will vary enormously, according to the nature of the substances under treatment; most emulsions can be produced with a single flow-through operation, at a rate of 2 tons per hour. Solids will take longer, and in the case of colloidal sulphur the output would be from 220 lbs. to 440 lbs. per hour, according to feed. Power required varies from 8 h.p. for emulsions to 10 to 15 h.p. for pigments in oil and water. In the case of very heavy substances such as barytes, 20 h.p. is required.

Mills which effect disintegration and mixing solely by shearing action between surfaces are many. A well-known example of this type is the Premier mill,[1] handled in this country by Messrs. Burt, Boulton, and Haywood, of Silvertown, and illustrated in Fig. 9. This consists of a casing in which a rotor, R, shaped like a truncated cone, revolves in close proximity to a similarly

[1] "The Premier and other Disintegrating Mills and their Applications in Industry," by F. J. E. China.

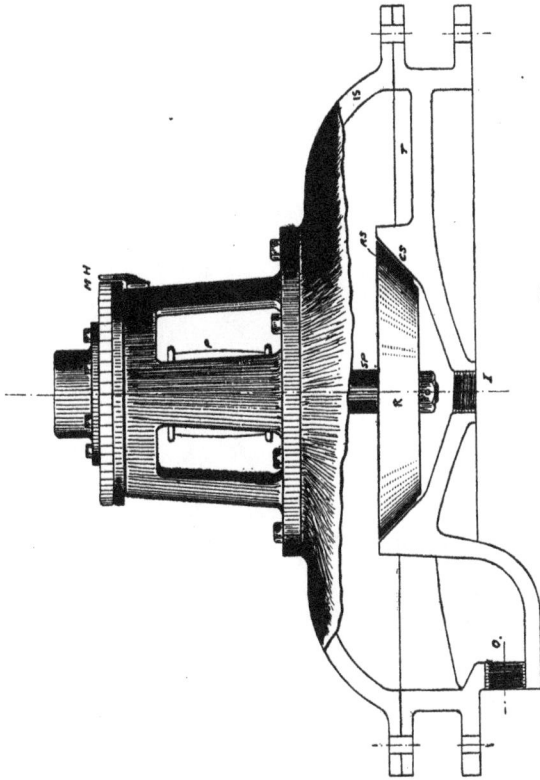

FIG. 9.—THE PREMIER MILL.

shaped surface, CS. These two members must on no account touch, but may be set with a clearance varying between three-thousandths and two- or three-hundredths of an inch (according to the material being dealt with), by means of a micrometer head, MH. The materials, liquids, or liquids with solids in suspension, are fed in through the opening I, and are compelled to pass between the fixed and rotating surfaces. At the high speed (1000 to 5000 revolutions per minute), the materials are completely disintegrated by the powerful hydraulic forces brought into being. Very little head of liquid is required to feed the mill, owing to the centrifugal action of the rotor drawing it through. The mixed material leaves by the opening O.

The rotor, of high-grade steel, is machined perfectly smooth, and accurately balanced. The angle of the cone is usually 45°, but may be made smaller or larger to suit special circumstances. The standard 15-inch mill requires 25 h.p., and rotates at 3500 revolutions per minute. The output varies according to the viscosity and other properties of the materials being dealt with, being 25/35 gallons per hour for viscous materials such as enamels, up to 1000 gallons per hour for water emulsions.

Another type of machine depending upon the shearing forces between fixed and moving surfaces is the Hurrell homogeniser, invented and made by G. C. Hurrell at the

Sun Lane Engineering Works, Blackheath. Writing in the *Chemical Age* of April 26th, 1924, G. C. Hurrell describes two types working on the above principle.

(a) Those in which the liquid is propelled through the gap by the centrifugal forces acting upon the material forming the film. In these the surfaces are in the form of discs, or truncated cones diverging from the ingress end as in the case of the Premier mill.

(b) Those in which the surfaces confining the film are so shaped that the film does not tend to evacuate the space by its own inertia, or in which it even tends to work back into the feed space.

The Hurrell homogeniser is of the second type, and is illustrated in Fig. 10. The rotor, B, contains radial ducts, F. Liquid fed in through D enters these ducts and is driven outwards by centrifugal force, and enters the film gap, G, between the casing and the rotor. The solid portions of the rotor between the ducts act as impellers, and thus cause the liquid to exert pressure at the outer end of the film gaps. This pressure, and consequently the rate of flow through the machine, will depend upon the rate of feed.

The advantage of this system lies in the fact that it is thus possible to vary the

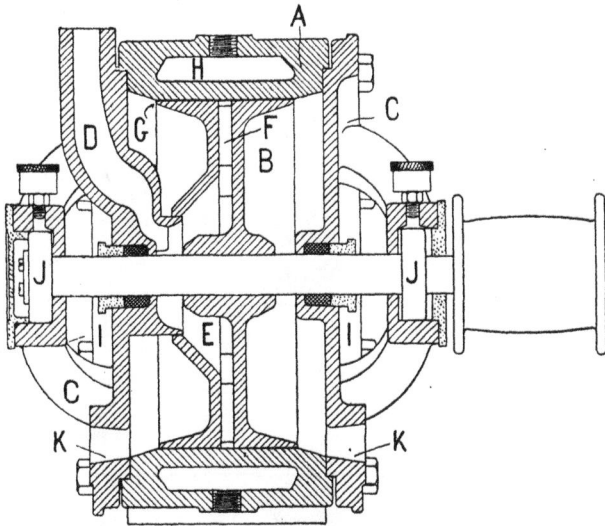

TYPE A. STANDARD MACHINE.

FIG. 10.—THE HURRELL "HOMOGENISER."

3

treatment according to the class of material by simply varying the feed. A material that requires much work to be done upon it is fed in slowly, and one that requires less, more rapidly.

With machines of type (a) the material being fed in at or near the axis of rotation, the film of material is probably here a solid stream, but as it moves outwards where linear speed and area are greater it tends to break up and form voids between the masses, and mixing is consequently not so effective.

The Hurrell homogeniser has a further advantage in that, as the liquid escapes upon both sides of the rotor, the hydrostatic forces are exactly balanced longitudinally, and consequently there is no end-thrust on the bearings. This type is suitable for continuous work on the same class of material. For dealing with different classes of material the same firm make a machine with slightly tapered rotor; this taper is not sufficient for the film to be affected by centrifugal force, but sufficient to enable the thickness of the film to be varied according to the nature of the material being dealt with.

The homogeniser rotates at speeds varying from 3000 up to 8000 revolutions per minute, the latter with a 20-inch rotor giving a linear speed of 40,000 feet per minute. The film gap is two-thousandths of an inch wide in the standard machine, but with the adjustable rotor this may be varied from one-thousandth upwards.

A machine invented by W. Ostermann, of Oschersleben,[1] has a horizontal conical rotor set at an acute angle. Both rotor and stator are provided with straight or helical grooves to increase skin friction. Helical blades are provided at the feed and discharge ends to force in and withdraw the liquid. In addition, a nozzle is fitted into the discharge pipe whereby hot, compressed air may be injected into the exit stream which thereby becomes atomised, and the mixture, striking against a baffle in a receiving chamber, falls down into the latter as a fine dry powder. This machine is interesting in that a dry mixture is produced. Plauson has also invented a dry grinding mill which could be used to produce dry mixtures of colloidal character; in this case, however, no liquid dispersion medium is employed.

Another type of mill employs two discs set in close proximity to one another and each rotated in the opposite direction. The advantage of this arrangement is that the speed of rotation need only be half that required in a machine with one fixed and one moving surface. One example of this is a machine with vertical shafts by H. O. Traun's Forschungs-laboratorium G.m.B.H., of Hamburg. The material is fed through

[1] International Patent Specification 216110.

the upper shaft (made hollow for the purpose), and discharged at the periphery of the discs.

Another is the super-mixer made by the Low Engineering Co. Type A of this machine consists of a pair of adjacent discs mounted on shafts in line, the whole being encased. The discs are made slightly conical, the concave faces facing one another; upon these faces are mounted rings of projections, and these rings register alternately with those on the opposite disc. At the peripheries of the discs the projections are in the form of complete rings, the clearances being extremely narrow.

There is a hollow space, triangular in section, at the centre of the discs, into which the liquid is fed from both sides and whence it is thrown outward by centrifugal force, and, passing between the discs, is subjected to a continuous beating and shearing action by the projections and rings; a micrometer adjustment is provided for varying the film gap.

The discs are driven in opposite directions by means of pulleys mounted on each side of the casing. The machine is capable of treating from 30 to 100 gallons per hour, according to the viscosity of the liquid, using a motor of 10 h.p. The discs revolve at a relative speed of 6000 r.p.m.

Type C is a small, hand-operated machine containing three separate rotors. The largest is hollow, concentric with the casing and provided with internally projecting teeth, and carries a boss provided with teeth projecting outwards. Both these sets of teeth register with those upon two rotors mounted eccentrically, and are driven by gearing in the opposite direction to the large rotor at a speed rather more than three times as great.

The action of this machine is of a beating rather than a shearing character, and is similar to the Plauson mill in this respect, the difference lying in the fact that both beaters and anvils are moving in opposite directions, thus increasing their relative speeds.

As an emulsifier and intensive mixer for laboratory and domestic purposes, this machine is very useful indeed.

Messrs. J. Harrison Carter, Ltd., of Dunstable, have recently brought out a similar type of machine to the type A, with fixed and rotating rings; both these are provided with concentric grooves, which mesh with one another; in addition, each ring has a number of grooves tangential to its inner circumference which intersect the concentric grooves. The material is fed into the interior of the rings, whence it is whirled outwards by centrifugal force and disintegrated between them; the tangential grooves tend to bring the material back and to counteract the centrifugal force, throwing it outward. The discs may be set from one-thousandth to a few hundredths of an inch apart. It is claimed that this arrangement reduces the material to a finer degree than plain discs or rings.

Uses of Intensive Mills. — The uses of these mills are extremely numerous; in fact, there seems to be no end to their applications. Confining attention more especially to their use as mixers, we may mention the following.

Rubber Mixing.—It is essential in the preparation of rubber mixes to get them as homogeneous as possible. To this end the added compounds, sulphur, lime, zinc oxide, and fillers, such as carbon black, French chalk, must be in extremely finely-divided form.

Intensive grinding and mixing for this purpose can be more efficiently carried out in mills of the above character than by any other means.

Colloidal Fuel.—Suspensions of coal in fuel oil for feeding oil burners and for all purposes in which fuel oil can be used, can readily be produced. Also emulsions of petrol and water for internal combustion engines. An emulsion of this character containing 70% of petrol has been used successfully.

Lubricants.—Emulsions of oil in water, and suspensions of graphite and talcum powder in oil, may be made.

Creosote Emulsions.—The advantage of these preparations lies in the ability to use a much smaller percentage of creosote for preserving timber. When using neat creosote, in order that the preservative may penetrate to all parts liable to become exposed by cracks or otherwise, in sufficient concentration to be effective, the outer layers have to be over-impregnated. A creosote emulsion is just as effective, using a fraction of the quantity of creosote required for the same depth of penetration. Emulsions containing 30 to 50% of creosote may be used, and are so stable that they may be boiled or frozen without breaking down owing to aggregation of particles.

Pharmaceutical Products.—Such preparations as face creams, and the so-called liquid face powders, hair creams, disinfectant emulsions, emulsions of essential oils for perfumes, may all be quickly prepared. The Low Engineering Company Type C hand-operated mixer is very useful for this purpose, enabling retail druggists to make up their own preparations.

Foodstuffs.—Milk powder may quickly be ground up and mixed with water to the consistency of ordinary milk. Meat may be extracted in the cold for the preparation of meat extracts, and fruits may be rendered fluid. Flavouring essences may be prepared by emulsifying suitable essential oils in water. In addition to emulsifying and colloidalising, many of these machines actually sterilise the preparation, owing to the fact that the bulk of the particles is reduced to a size less than $0·5\mu$, with the result that any organisms larger than this are broken up and destroyed. As practically all known organisms are larger than this,[1] sterilisation is effected, an important point in connection with articles for human consumption.

Other Uses.—The manufacture of viscose for artificial silk involves the treatment of cellulose with caustic soda. In the usual method, considerable excess of the latter has to be used, but so intimate is the mixing effected by certain types of colloid mill, that the theoretical quantity is sufficient. In a similar manner, tar acids may be extracted from tar oils far more rapidly than by the ordinary processes, owing to the extreme intimacy of mixing.

Many pigments of a soft nature may be ground and mixed with oil in the preparation of enamels or printing inks, and a smooth, even mixture obtained without the wear and tear which usually characterises ordinary grinding and mixing machines for this purpose.

Many dyestuffs, which in the ordinary way are extremely difficult to mix with water, may easily be made up into permanent fine

[1] Average bacilli are 3 to 6μ in length and $0·5$ to 1μ in diameter ($1\mu = 0·001$ mm.).

suspensions, and inorganic pigments may in some cases be obtained in suspension sufficiently fine to be used in the dye-bath.

It should be explained that, in most cases, in order to ensure permanency of suspensions or emulsions prepared in colloid mills, it is necessary to add a small percentage (usually under 1%) of a stabiliser which is itself a colloid.   Examples are, such substances as gum, gelatine and soap; their function is, however, solely to stabilise the preparation, and they play no part in obtaining it, so do not lie within the scope of this book. Numerous other uses of colloid mills might be given, but as this subject is already being dealt with in another book of this series, there is no occasion to go farther.   The subject, from the point of view of mixing, however, is important and interesting, and it is for these reasons that this chapter has been included.

CHAPTER V

MIXING IN THE CEMENT AND BUILDING INDUSTRIES

THE manufacture of cement involves the calcination of a mixture of argillaceous and calcareous materials, *e.g.*, clay and chalk, shale and limestone. The mixture is burnt to clinker in kilns of rotary or other type. As it is not actually fused, but merely sintered, complete chemical reaction between the constituents can only be brought about by the most intimate mixing. The ideal method of making cement would involve the actual fusion of a mix containing approximately 70% of lime, in a kiln of the blast-furnace pattern. Such a process is quite sound and practical from a scientific point of view, but owing to the high temperature (over 1700° C.) required for the fusion it is not economically sound.[1] If it were, however, the question of mixing would not be of nearly so great an importance. A comparatively rough mix of the constituents in the right proportions which, on descending into the fusion zone, would quickly react together in the liquid state, would give a perfect cement. As this process is not at present practicable, however, it is essential to mix the raw materials as intimately as possible, so that, at the clinkering temperature (1200° C.), reaction

[1] See " Cement," by Bertram Blount, Longmans, Green & Co.

will go to completion, there will be no portion of either constituent left over, and a cement which does not require ageing or slaking and can be filled straight into bags from the grinding plant may result.

There are two processes in vogue for mixing the raw materials—namely, wet and dry. According to the principles enunciated in Chapter I, this involves fine grinding, and as both raw materials are in the case of chalk and clay of approximately the same degree of hardness we find that, to a great extent, grinding and mixing are effected in the same machine.

Wet Process.—Owing to the extreme intimacy of mixing possible with this process, it is, in general, the most preferred. The raw materials, as brought from the quarries, are weighed out and dumped into a type of mixer known as a wash-mill. This consists of a large circular pit lined with concrete or brick, and containing in the centre a massive pillar of similar material; a sort of annular pit or moat is thereby produced. The pillar serves to support a vertical shaft carrying horizontal arms from which rakes or harrows are suspended by means of chains. The vertical shaft is made to revolve at from 20 to 30 revolutions per minute by means of crown and bevel gearing. The mill is filled with water up to a point, the agitator set in motion and the raw materials are dumped in. The rakes, swinging round, strike and disintegrate the lumps of material, but not being rigidly fixed and merely depending from the arms, they are not damaged by contact with any lumps of material too hard to disintegrate; such lumps drop to the bottom and in time accumulate. For this reason provision is often made for raising the harrows so as to clear accumulated *débris*, until this becomes so great as to materially reduce the capacity of the mill, when it is cleaned out. A thick slurry of approximately correct composition is thereby obtained. The principle is not unlike that of a levigating mill for preparing fine pigments. The ordinary type of wash-mill usually has four arms carried on the central shaft, and may have a capacity of 50 tons of raw material per hour.

Wash-mills are usually of the batch mixing type, and the batch, when thoroughly mixed, is run out through screens, which stop any coarse material, to dosage tanks. In these, any inaccuracy of composition is corrected by the addition of the appropriate quantity of either raw material. This is a large tank similar to the wash-mill, but sometimes oval in shape, and contains one or more agitators to keep the slurry uniform and prevent settling. These agitators are similar to the wash-mill agitator, but as there are no lumps to break up here, the rakes are rigidly fixed to the arms instead of hanging by chains. An elevation of a wash-mill is shown in Fig. 11.

From the dosage tank the slurry is passed to the storage tanks, also containing agitators of similar pattern to prevent settling.

Sometimes the slurry is pumped from the wash-mill to a centrifugal separator lined with wire netting and known as a

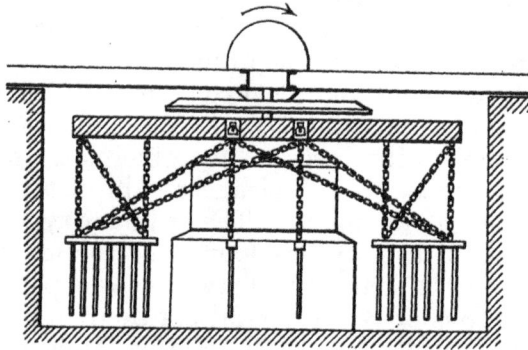

FIG. 11.—CEMENT WASH MILL.
Adapted from Blount's "Cement."

"trix," whereby the coarse material is retained and returned to the wash-mill. From thence the slurry, now fairly fine, but still not sufficiently so, passes to a tube mill, where it is wet ground and more intimately mixed. It is now ready for storage and for the kilns.

The extreme importance of intimate mixing in the manufacture of cement is so great that the saying has arisen that " cement is made in the wash-mill." Although the actual formation of cement takes place in the kiln, yet unless the preliminary preparation of the raw material by mixing is properly carried out, no amount of care in burning will produce a satisfactory product.

The afore-mentioned process is for mixing soft materials such as clay and chalk, and is the system employed at the Portland Cement Works, on the banks of the Thames, where the raw materials are mud from the river bottom and chalk from the quarries along the banks. In dealing with hard materials such as shale and limestone, these must previously be crushed and ground, as the harrows of the wash-mill would have but little effect on them. This is effected in a series of machines starting with preliminary breakers, through preliminary grinders, such as crushing rolls, to fine grinders of the ball and tube mill variety. When the wet process is in use, the fine material is fed to a mixer of the wash-mill type, where slurry containing about 40% of water is produced.

The wet process has the following advantages. More intimate mixing is possible in a liquid medium and a slurry produced which can be handled by pumping or gravity, and readily fed to the kilns. Less power is also absorbed in wet mixing. The disadvantage lies in the extra fuel required for driving off the 40% of water. Generally speaking, however, the wet process is commoner than the dry, and is preferred for soft materials.

The dry process is usually employed for hard materials such as limestone and shale. In order to incorporate them properly, it is essential that they should be thoroughly dried before mixing. The limestone as quarried is fairly dry, but the shale varies considerably in its content of water; both, however, are put separately through a dryer usually of the rotary type and then crushed to approximately 1-inch cubes. The correct proportions are then roughly blended in a simple revolving cylinder; a tube mill without the balls, or in which arms have been substituted, may be employed. This mixing is very rough, owing to the large size of the pieces; it is completed in the grinding mills, which may be any of the usual types. The ball and tube mill installation has the advantage that owing to its large capacity any errors in composition are corrected by "evening out" of the materials. The ball mill as used is sometimes a modified type known as a "Kominor." It acts as a batch mixer as well as a grinder, as the plates are solid and not provided with holes, which tend, in time, to close up owing to the continual hammering action of the balls. The material is finished in a tube mill, which completes the grinding and acts as a continuous mixer, the mix, ready for the kiln, being delivered through the hollow discharge trunnion (see Chapter III). The material must be quite dry before being fed to the grinding plant, especially if this be of the above type, as otherwise clogging will take place, the balls tending to become coated with the ground material.

Various other types of grinding mills are used, such as mills of the ring-roll type, giant griffin, etc., but for details of these the reader is referred to works on grinding. The above have only been referred to on account of their function as mixers.

The dry process, although saving the fuel used in drying the mix in the kiln, is not so common as the wet, on account of the difficulty of properly mixing the materials. Against this saving must also be put the extra cost of grinding and mixing the materials. The importance of proper mixing of the raw materials cannot be over-emphasised if a satisfactory product is to be obtained.

Concrete.—The mixing of concrete naturally follows that of cement. This process involves the mixing of the ground cement clinker with some kind of aggregate such as sand for smooth surfaces like floors, coarser materials such as granite, limestone and brick rubble for foundations and structures. Small quantities are often mixed by hand, but for satisfactory work a mixer is better.   To obtain a perfect concrete every particle of aggregate must be evenly coated with cement, and simple mixing by hand does not effect this properly.   A properly made concrete mix should contain sufficient water to make it flow like cream; in addition, all air should be driven out, so that, on setting, a dense, hard body is produced. The mixer must therefore be designed with these objects in view.

A simple type suitable for hand operation consists of a hollow drum fixed in an inclined position for mixing, but capable of being tipped for the purpose of emptying.   Sometimes, in order to increase the capacity, the drum is made in the form of the frustum of a cone, the smaller open end being uppermost.   A series of shelves is fixed all round the inside; [1] these shelves pick up the materials and carry them round until the former are sufficiently high to cause the material picked up to fall back into the trough.   The materials are thereby continually turned over until they become intimately mixed.

An example of this type is the London Gem concrete mixer made by the London Concrete Machine Co., Ltd., of London (Canada), and handled in this country by Messrs. George Waller and Son, Ltd.   It may be worked by hand or power, and

---

[1] These shelves are frequently arranged in helicoidal fashion to increase their effective length.

complete portable units with small ($1\frac{1}{2}$ h.p.) petrol engines are made. The capacity is about $2\frac{1}{2}$ cubic feet of mixed concrete and the output about 20 cubic yards per day.

Another type of concrete mixer is the chain-spade type. This is a horizontal shaft mixer. The shaft, which is square in section, has arms bolted to it carrying at their extremities spades which dig up and turn over the materials. In addition, the ends of adjacent arms are joined by short lengths of chains which sweep the bottom of the container and keep it clean. The latter is made semi-circular in section and is provided with tipping gear for emptying. The batch is in full view during mixing, so that water may be added gradually until the whole mass is suitably tempered.

Examples of these mixers are the " London " chain spade mixer, made by the above firm, and the " Winget." A hopper may be provided above the container with a sliding door for feeding the materials. Instead of spades the ends of the arms may be sharpened to a kind of blunt knife-blade which acts in a similar manner.

These mixers have a capacity of from 5 to 7 cubic feet and require from 4 to 6 h.p. They are particularly suitable for mixing semi-wet concrete for the manufacture of concrete shapes such as blocks

and tiles. A smaller proportion of water
must be used for this purpose than for
ordinary concrete in order that the shapes
may not deform under their own weight
before setting. Owing to its stiffness, such
a mix cannot as a rule be made in the
ordinary type of mixer. The chain-spade
type gives a powerful, positive action and
is thus eminently suited for this purpose.

This type may also be used for mixing
mortar in place of the usual pan mill with
one or two rollers (see Chapter III). It is
also suitable for mixing wall plaster and
hair mortar.

A third type, also made by the London
Concrete Machinery Co., is a drum mixer.
The drum revolves on a horizontal axis,
upon bearing wheels, being driven by means
of an external toothed ring. A vertical
section of the drum is shown in Fig. 12.
It is closed at both ends save for a slight
hole, A, through which observation may
be made of the progress of the mixing. It
is provided at the other end with a combined feed and discharge chute, B, which
remains stationary during mixing. It is
shown in the diagram in the charging
position, the discharge position being indicated by the dotted lines.

The action is as follows :—

The feed chute is set in the charging
position, the materials are placed in the
hopper, and on lifting the door the whole

batch drops into the drum. It is caught
by the buckets, C, as the drum rotates
and carried to the top, whence it is spilled
downwards upon the convex side of the
chute, which can be reversed for this purpose. This action spreads the material and

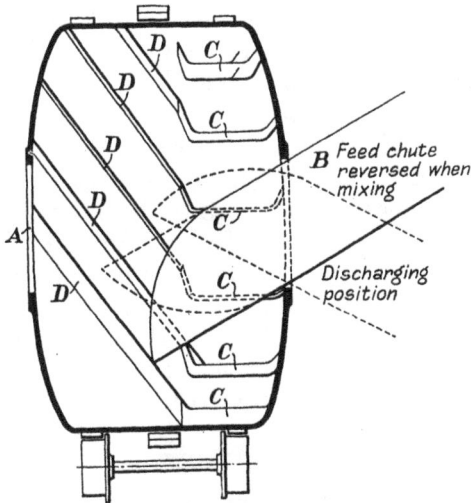

Fig. 12.—Drum Concrete Mixer.

forces all the air out of it; in addition, it
spreads it in the reverse direction to the
action of the blades, D. These blades slice
the material, and the part which is caught
by them slips down into the buckets, to
be again carried up to the top and spilled
upon the reversed chute again.

The combined rolling, spilling, spraying

and slicing actions effect an extremely thorough mixing in a few minutes. Discharge is effected by moving the chute to the position indicated by the dotted lines, when the material dropped from the buckets falls into it and slides out.

This is a highly efficient mixer and produces a dense, homogeneous mix free from air pockets, in which every particle of aggregate is coated with cement. The drum is self-cleaning, as the force with which the materials are tossed about effects a scouring action upon the internal surfaces, thereby keeping them smooth and polished. Practically all concrete mixers are of the batch type, continuous mixing not finding favour with engineers in this country. This is probably due to the difficulty of accurately measuring the different varieties of aggregate, varying from sharp sand to coarse stone.

Tar Macadam.—Tar macadam and concrete are materials not dissimilar, and so, for the sake of completeness, a few details are here included, although strictly it cannot be classed as a chemical industry. As is well known, it is produced by mixing a suitable aggregate, such as granite or other stone, with tar or bitumen in a molten condition. The latter, on cooling, sets, and binds the whole into a mass. As with concrete, it is important to coat every particle of the aggregate with the binder. In order to

ensure a thoroughly hard mass, the aggregate must be suitably graded into coarse,
medium or fine material, the whole being
cemented together with just sufficient tar
to fill the voids and no more. By suitable
grading these voids are reduced to a minimum, the medium material filling those
between the coarse material, and the fine
those between the medium. The tar com-

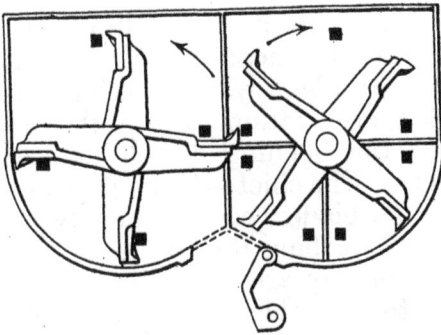

FIG. 13.—FAWCETT'S TAR MACADAM MIXER.

pletes the filling. Intimate mixing is,
therefore, necessary to obtain a satisfactory
result.

Fig. 13 shows a suitable mixer in longitudinal and cross section made by Thomas
C. Fawcett, Ltd. It is of the double shaft,
horizontal type. The shafts revolve in
opposite directions, as shown by the arrows;
they carry arms to which are bolted steel
paddle plates, A. The aggregate, granite,
slag or other material, is dropped in after

having been dried and heated in a suitable heater, and the hot tar added by means of a pipe mounted over the mixer. The material is raised in the centre by the paddle plates, turned over and dropped into the container, to be again lifted, and kneaded in such a way as to produce a thorough mixing of the constituents. Discharge is effected through the door, B, which, being the full width of the container, allows the whole batch to drop out into a wagon placed beneath. This is facilitated by the paddles, which, owing to their direction of rotation, sweep the material towards the opening and lift it up over the same. The chief wearing parts, namely, the paddle blades, can be quickly renewed without disturbing the shafts.

These mixers are made in sizes varying from 5 to 27 cubic feet, *i.e.*, about 30 to 200 tons per day, and require from 5 to 20 h.p.

Concrete and tar macadam mixing lie somewhat outside the scope of this book, but have been included as interesting examples of mixing, and as a complement to the subject of cement.

CHAPTER VI

MIXING IN THE CERAMIC INDUSTRY

THE ceramic industry, including as it does the manufacture of bricks, especially fireclay, silica and other refractories, stoneware, glass, china and porcelain, is an extremely important and wide one. Although not strictly a chemical industry, yet chemistry enters into it to such an extent as to fully justify its inclusion here. Numerous and peculiar problems of mixing present themselves, but it will only be possible to deal with a few of them here.

Fireclay Products.—The manufacture of fireclay bricks and blocks involves the mixing and blending of the ground clay, usually with the addition of a percentage of calcined material, or " grog," as it is termed, which may amount to as much as 60%. This grog is added to reduce shrinkage of the blocks in firing, to give increased porosity and hence greater resistance to sudden changes of temperature, and to reinforce the material in the same way as cement is reinforced with an aggregate in making concrete. For the last-named purpose the grog must not be too fine, and in order to be effective it must be evenly distributed throughout the mass.

Mixing is usually effected in large pan mills, which, when used as continuous mixers, have an output up to 8 tons per hour of ground and mixed material; for

this purpose, the pan bottoms are perforated with holes, usually rectangular, as
described in Chapter III. The materials,
raw clay and grog, are added, the quantity
usually being very roughly gauged by
barrow-loads. The raw clay may be thrown
in as large lumps, as, being a soft material,
it is quickly crushed by the rolls. The
grog is usually broken to start with in
stone-breakers of the swinging-jaw or
rotary pattern, down to pieces the size
of the fist. Continuous mixing in perforated pans is effected in the dry state
and the dry mix falls through the holes in
the pan down into the boot of an elevator
below, whence it is elevated and distributed
over screens usually not finer than 4 meshes
to the linear inch, and often coarser, say
a $\frac{3}{8}$-inch mesh, or coarser still for some
types of goods. The tailings are returned
to the pan for further grinding. Screens
with round holes in place of wire mesh are
sometimes used on account of their greater
strength and durability, but are not so
efficient. The screened material has now
to be wetted in order to render it plastic.
This is by no means so simple a matter as
might at first be supposed. The material
carried by a band conveyor is sprinkled
with water by means of a perforated pipe
fixed over the band. The raw clay absorbs
a considerable quantity of water and the
damp mix is finally discharged into a deep

bin to " age." This ageing process is extremely important, as, if not properly carried out, the mix will not be evenly plastic, but short and " rawcey " in places, and tend to stick in the pug mill.

The period of ageing varies from two or three days up to six months, according to the character of the mix. Materials containing little or no grog, such as stoneware clays and the like, are often aged for long periods, in some cases even for years.[1] Heavily grogged or highly siliceous materials such as the argillaceous sands of the High Peak district of Derbyshire, require only a few days' ageing, owing to the small proportion of clay they contain. This ageing or soaking process is maintained because it is really an automatic mixing which takes place slowly throughout the mass. Its exact nature is not known, but the water probably percolates throughout the mass by capillary action and so becomes more evenly disseminated. It is sometimes thought that bacterial action plays a part, and for this reason clays are sometimes aged in the dark. The principal process, however, and the one which renders the clay easy to work, is simply soaking.

The ideal to be aimed at, although hardly

[1] The clay used for making pottery in China in olden days is said to have been aged for 100 years ! This seems amazing, but is not improbable, time being no factor among Oriental peoples.

attained, is to coat every particle of material with a film of water. This may be hastened by wet mixing in a mixer, preferably of the horizontal shaft type. Fawcett's double-shafted clay-mixer is eminently suited for this purpose (see Chapter III, Horizontal Mixers).

The dry or semi-dry mass is passed through the mixer and water from a sprinkler pipe added. The two shafts, revolving in opposite directions at different speeds, knead and turn the mass over, thereby thoroughly incorporating the water with the clay and producing a homogeneous mass.

Mixtures which previously were too short and "rawcey" to pass through the pug mill, such as highly siliceous materials (over 90% silica), can often be brought down to a smooth, even paste, which slips easily out of the die of the pug (see Pug Mills below). In some cases the materials are ground and mixed wet in a pan having a solid, revolving bottom and stationary sides, or in a stationary pan with travelling rolls. This is a batch mixer, and the batch may be discharged through a door in the side or bottom of the pan. Where wet mixing is employed there is no necessity to dry the clay before use.

After grinding, mixing and ageing in the above manner, the material is next pugged. It is elevated and passed between heavy iron rolls; these squeeze the material, incorporating the water more intimately and rendering the mass more homogeneous. One or more pairs of rolls may be used; in the latter case the pairs are placed one above the other, the material passing between each pair in succession. The distance between the rolls is adjustable according to the fineness of the material being handled, but their function is solely one of mixing, and not crushing.

Driving is effected by means of pinion wheels mounted on the ends of the roll shafts and meshing together. Each pinion may have the same number of teeth so as to drive the rolls at the same speed, but to reduce wear upon the teeth it is not uncommon for one pinion to have one more (or one less) tooth than the other.

As a rule, the surfaces of the rolls are smooth, and the rolls themselves cylindrical. But occasionally the surfaces are fluted, or have longitudinal grooves cut in them to increase the surface and thereby enhance the mixing. The rolls are sometimes made conical for the same purpose.

After passing through the rolls the material falls into the pug. This is a cylinder containing a shaft having blades upon it arranged as an interrupted worm, which force the stiff paste along and extrude it through a die, after which it is cut off at intervals, and the lumps, or " clots," as they are termed, go to machines to be made up into shapes.

Pugs are of two kinds, vertical and horizontal. The latter is the more popular, as it will deal with stiffer bodies. The function of the pug is principally conveying and extrusion, but as additional water is frequently added in it, in order to get the right degree of plasticity, it has a certain mixing function. It also compresses the body, thus rendering it more dense and tough. The mixing function is sometimes enhanced by fitting stationary counter knives to prevent rotation of the mass; or a pair of shafts rotating in opposite directions are sometimes employed for the same purpose.

Pugging improves all plastic materials by rendering them more dense and homogeneous. With non-plastic materials, such as sands and ground silica rock used in the manufacture of silica bricks, it is not practicable unless a sufficient proportion of water be added to produce a slop. Any attempt to do so causes the mass to jamb inside the body of the pug, with consequent breaking of the knives. The same trouble sometimes occurs with highly siliceous materials such as argillaceous sands, but may often be overcome by kneading the mass with water in a horizontal mixer before passing it through the pug, as previously described. Where no clay is present even this is not practicable.

Silica brick mixes of this character are prepared in solid-bottom pans. The rock is ground for a certain period, 10 or 15 minutes, dry; milk of lime is then run in during a further period of 5 minutes; this serves to incorporate the lime throughout the mass. The batch is removed from the pan by means of a scoop or sliding door in the bottom and made up at once by hand, no period of ageing being required in this case. The mass may be compared to a clay mix composed entirely of grog if such a mix were possible. The shapes hold together when dried by the setting of the lime in the same way as mortar. Owing to the comparative coarseness of the material compared with clay, the water quickly percolates through, so that thorough mixing does not present the same problem in silica brick manufacture as it does in fireclay.

Owing to the importance of getting the correct proportion of lime in the mix, silica brick mixes are invariably made in batches. 2 to 3% of actual lime is usually added and the correct proportion must be maintained if bricks of uniform composition and refractoriness are to be obtained. With fireclay bodies containing 40 or 50% of grog, however, slight variations in proportion of grog and water do not produce the same effect, and consequently continuous mixing is preferred for large outputs.

A simple device for regulating the addi-

tion of water to a mix may be described.
The device is illustrated in Fig. 14. The
mixer, A, is of the horizontal type. The
clay or other material is fed from the
vertical chute, B, through an orifice the
size of which may be varied. Water is
supplied from the pipe, C, to a per-

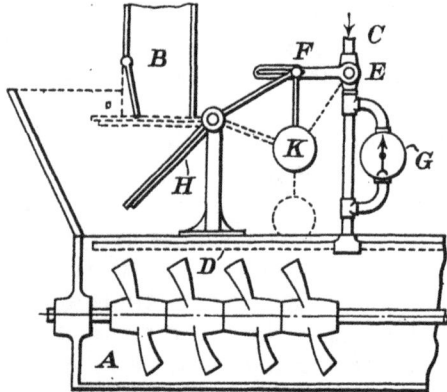

FIG. 14.—CLAY MIXER.
Adapted from the *British Clayworker*.

forated pipe, D, the feed being regulated
by the tap, E, actuated by the lever, F.
The exact amount of water passing can be
measured by the meter, G. The pivoted
plate, H, controls the supply of clay, and
this can be regulated by means of the
balance weight, K, which can be moved
along the lever arm, F. The plate controls
the movement of the lever, F, as well. If
the supply of clay diminishes, the left half

of the plate rises, depressing the right half
and reducing the supply of water. If the
clay feed ceases altogether, further movement of the plate completely cuts off the
supply of water.[1] The device is a simple
one and can easily be fitted up in the
average works. Adjustment is quickly
effected by moving K along the lever arm.
The water meter is not essential, but is a
great convenience both for regulation of
the supply and operation.

Turning now to the finer ceramic industries such as the manufacture of bodies for
stoneware, china and porcelain, we find
that a much higher degree of intimacy in
mixing is required than for coarser bodies
such as bricks and blocks. The materials
are much more finely ground and incorporated.

Stoneware is the name given to bodies
made from clays burnt to such a degree
that they become vitrified, practically the
whole matrix being converted into a glass,
so that the whole body becomes non-
porous. The highest degree of vitrification
is reached in chemical stoneware, which has
to stand the action of hot, concentrated
acids and must therefore be quite imper-
vious, otherwise penetration and disintegration quickly result. The bodies are usually
made up by mixing the clays and water in
a mixer not unlike that in which bread is

[1] See *British Clayworker*, June 1924.

4

kneaded. Powerful blades knead and turn over the whole mass until a stiff paste free from air bubbles is produced. This is then left to age, or " sour," as it is termed, for periods varying from six months to two years before being made up.

China and porcelain are prepared from a variety of materials, but the basis is, of course, china clay and ball clay with the addition of flint and felspar, bone ash, or whiting; numerous other materials are used, but need not be dealt with here.

The method of incorporating these materials is as follows. The ball and china clays are placed in a type of mixer known as a blunger. This is not unlike a cement wash-mill, and consists of a circular wooden vat having one or more vertical shafts each carrying two horizontal wooden arms from which depend harrows. As there are no hard particles to be broken up, these harrows are rigidly fixed to the arms.

Horizontal shaft blungers are sometimes used. The shaft carries a series of discs with flanged peripheries; longitudinal battens are bolted to the flanges and carry in their turn wooden beaters which beat the material into a slip. The whole is contained in a drum which may also be made to rotate in the reverse direction to the shaft, thus doubling their relative speeds and preventing the settling out of heavy materials. The clay and water are accurately proportioned to give a slip containing a definite quantity of material per pint, so that it may be measured into a mixing vat by depth. The flint and stone are likewise converted into a slip of definite composition and the correct quantity is run into mixing vats with the clay slip. Sometimes the slip is passed through a sieve or " lawn " of fine mesh (130 meshes to the linear inch) to separate coarse particles and give a smooth cream.

American practice consists in preparing the slip in a ball mill. The materials are ground with water to a fine cream.

The slip is next passed through a filter-press and the resulting cakes are again blunged with water to which some deflocculating agent, (usually sodium carbonate or sodium silicate), has been added to increase the density of the slip. By this means a slip weighing 36 or 37 ounces per pint, and therefore containing less water, can be produced, whereas with water only, 27 ounces per pint would be the heaviest slip which would flow.

The shapes, such as lavatory basins and other sanitary ware, are made up by running the slip into the plaster moulds, which quickly absorb the water, causing the body to shrink away from the mould, so that it may easily be removed.

The above is the method for large and complicated shapes; for small pottery bodies the pressed cake may be pugged and made up on the potter's wheel, or by other means, as a paste.

The preparation of ceramic bodies and glazes is such a wide subject that it is impossible to do more than simply indicate the methods in use for mixing and blending the constituents so as to produce a homogeneous mass which will vitrify evenly and produce a dense, hard body. The method of blunging with a relatively large amount of water illustrates the advantage of wet mixing over dry. The degree of homogeneity thereby obtained is retained on de-watering, and the resulting paste may then be readily converted to the desired consistency for casting or moulding.

Before leaving the subject of mixing in the ceramic industry, reference may advantageously be made to a method in vogue in America and recently introduced into this country for the mixing of materials for brick-making, although it might be employed in a variety of industries for mixing materials in bulk, provided these did not suffer in quality through rubbing or squeezing (see Mixing in the Fertiliser Industry, Chapter VII).

The screened materials are delivered in succession on to the upper run of a steel belt conveyor, in the correct proportions, and thus form super-imposed layers. After the last feeding point a series of ploughs turns the material over and roughly mixes it; at the same time, water is sprayed upon it by compressed air through a series of atomisers. The wet mix is finally ploughed off the upper run on to the lower, whence it is delivered to the different presses.[1] This arrangement, therefore, combines the functions of conveying, mixing and moistening. Steel belts are coming increasingly into use on account of their wear-resisting qualities, and provided they are run at the correct speed (as a rule not more than 200 feet per minute) over pulleys not less than 40 inches in diameter, they are quite satisfactory.

[1] See *Jour. Amer. Cer. Soc.*, July 1924. This plant is made by Steel Belt Conveyors, Ltd., Birmingham.

CHAPTER VII

MIXING IN THE FERTILISER INDUSTRY

THE manufacture of fertilisers involves the intimate mixing of materials which in some cases require special treatment, as they cannot be handled in the ordinary types of machines. Moreover, efficient mixing is highly important, first because all fertilisers are required by law to contain definite quantities of each important constituent, namely, phosphoric acid, nitrogen and potash; and these quantities have to be stated upon the containers, and are subject to analysis by the Official Agricultural Analysts. Secondly, although by no means less important, unless the mixing is efficient so as to produce compound fertilisers which are homogeneous, when applied to the land some portions of the latter would receive insufficient of one constituent and too much of another, and the effect upon the resulting crops would be harmful either way.

The principal constituents of most fertilisers are : phosphoric acid in the form of calcium phosphate; nitrogen in the form of ammonium salts such as sulphate, nitrate or chloride, nitrate of soda, organic matter, or cyanamide (the last has a very limited use); and potash in the form of chloride or sulphate.

In order to render the phosphoric acid soluble and therefore assimilable by plants, it is usually converted into acid calcium phosphate by the action of either sulphuric acid in the manufacture of ordinary superphosphate, as it is termed, or by the action of phosphoric acid in the manufacture of double superphosphate.

This process, which is, of course, a wet mixing, is carried out in simple types of mixers of the vertical or horizontal type. In the former case, the mixer usually takes the form of a cylindrical container with a hemispherical bottom, made in special, acid-resisting, cast iron. It is provided with discharge doors opening outwards at the bottom. The agitator consists of a vertical spindle carrying mixing blades of helicoidal shape arranged so as to agitate and circulate the semi-liquid mass. Such a machine is illustrated in Fig. 15.

It is customary to fix the mixer above two or more superphosphate chambers or " dens," as they are called, in such a manner that discharge may be effected into any one of the dens by opening the appropriate door in the bottom.

The process of mixing or " dissolving," as it is termed, is effected as follows. The agitator is set in motion, and the acid and raw phosphate, in the form of mineral phosphate or bone meal, are added simultaneously. Agitation proceeds for a definite period, usually two minutes, found by experiment to be the minimum required to produce a homogeneous cream. The appropriate door is then opened and the mix discharged by gravity into the den; the door is then closed, a further batch run in, mixed and discharged as before; the cycle of operations is continued until the den is full, when the next is filled, and so on.

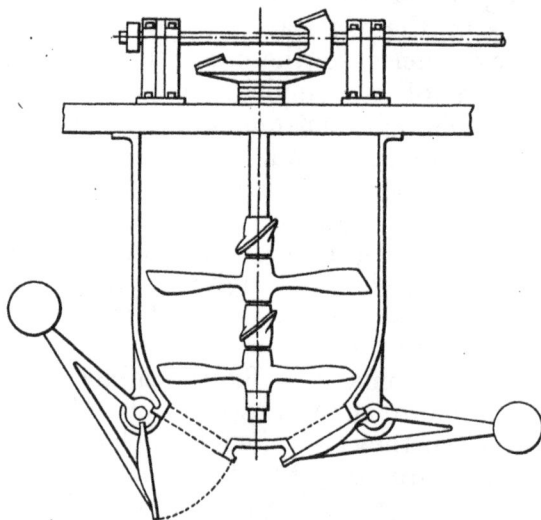

FIG. 15.—VERTICAL MIXER.

The mixing is to ensure intimate contact between the acid and the phosphate, the chemical action being completed in the den. The raw phosphate is previously ground to pass a 60-mesh sieve. The mix sets in the den to a soft, dry mass readily broken down with picks or by excavating machinery.

In some works horizontal mixers are used.
These are provided with a door at each end
for discharge. The mixer is fixed across the
top of two dens. The horizontal shaft
carries an open, helicoidal worm. This
agitator having been set in motion, the acid
and phosphate are introduced. The worm
forces the mass first towards the exit door,
which is, of course, closed; the direction of
rotation is then reversed and the mass forced
to the other end; the discharge door is
then opened, and the shaft being again
reversed, the worm forces the mass out
into the den below. This arrangement,
although giving a slightly quicker discharge,
lacks the simplicity of the vertical mixer,
which is kept continuously rotating in the
same direction. Unless somewhat complicated, automatic reversing gear is employed,
another attendant is required to reverse the
shaft at the appropriate intervals. Moreover, the strain put upon the spindle by the
continual reversal of direction necessitates
it being made stronger than would otherwise
be necessary. In the most modern factories
the whole process of dissolving is carried out
by one man. The phosphate is brought by
a conveyor to the mixing loft and automatically weighed; a magnetic device comes into
action as soon as the scale tips, whereby the
supply is cut off; the acid is controlled in
similar manner, a far more satisfactory
arrangement than the old one, in which the

4*

attendant had to cut off the flow when the level reached a mark in an open measuring tank; a very slight error of only half an inch or so made a considerable difference in the charge of acid to the mixer.

Continuous dissolving devices whereby the acid and phosphate are measured continuously, and flow into one end of a horizontal mixer-conveyor to be discharged at the other end into the den, have been evolved, but have not come generally into favour, at any rate in this country. Such arrangements would undoubtedly effect a marked saving in time, but the difficulty in measuring a solid, even when in a finely-divided condition, with the accuracy necessary has hitherto prevented the general application of such a scheme.

When we come to the manufacture of such a fertiliser as dissolved bone compound (called in the trade D.B.C.) intermittent batch mixing is the only practicable way. This material is prepared from raw phosphate and shoddy (wool waste), by treatment with sulphuric acid in the usual way. The shoddy is extremely bulky and fibrous, and would be practically impossible to measure continuously with any degree of accuracy. It is usually simply weighed out and introduced first into the mixer. The raw phosphate is put in on top of it together with the acid, and the whole thoroughly mixed. The material, a dark fibrous mass, takes about six months to mature in the heap, as solution of the shoddy by the acid proceeds very slowly.

The question of the order in which the materials are placed in a mixer is one of considerable importance in a large number of mixing problems. When mixing solids and liquids, it naturally stands to reason that the solid should not be put in first, as it would tend to cake on the bottom. The most effective way in most cases is to add the solid and liquid simultaneously. The words of Mr. L. M. G. Fraser, Chairman of the British Chemical Plant Manufacturers' Association, may be quoted with advantage here.

" Problems of mixing dry and wet materials can yet be studied with advantage, and a great deal of useful power saved by studying the rapidity of absorption of chemical reaction that is required to take place. Where the time factor comes in, large mixers with slow motion and less relative power may be advantageous, and with rapid solution smaller mixers and quick motion are advisable. I have known of cases where a very large amount of power was saved by the simple expedient of studying the order in which the materials were put into the mixer."

In the case of superphosphate and like substances, the question of chemical action in the mixer is not so important, as in any case the reaction completes itself on standing in the den, or in the heap in the case of D.B.C. But in order that this chemical reaction may proceed to the proper degree of completion, intimate incorporation in the mixer within a short space of time is highly important, and a careful consideration of the exact conditions of mixing is amply repaid.

Turning now to methods of mixing and blending compound fertilisers, we find a variety of special methods in use. Superphosphates and like substances are of a peculiar character. They must not be pressed or sheared to any extent, otherwise they tend to " paste," more especially when still hot from the den; on cooling, this paste sets to a hard, tough mass which cannot be ground in the usual way, as it merely pastes again, but has to be disintegrated by blows upon the unsupported material; even when treated in this manner it is often very difficult to pulverise, if in the form of small particles of the size of a pea. This is one of the chief reasons for the difficulty in designing suitable excavating mechanism for the dens, and why even to this day the pick and the shovel are still used to a considerable extent for this purpose.

Mixers for blending superphosphate with sulphate of ammonia and potash salts are designed with the object of avoiding, as far as possible, any grinding or squeezing action on the material. They are of both the batch and continuous types. An example of the former is the Sturtevant batch mixer made by the Sturtevant Engineering Co., Ltd., the principle of which is the same as that of the Drum concrete mixer described in Chapter V (Fig. 12).

It consists of a closed drum in which the same opening serves for the feed and discharge; thus there is only one opening to seal against dust. The end of the drum opposite the feed is made easily removable on the open-door principle, so that the interior is readily accessible for cleaning. This is an important point with fertilisers where frequent changes may be made in the mixings. Unless properly cleaned, the residue left in the mixer will cake in the manner peculiar to superphosphates. This residue will then be discharged with the next batch in the form of lumps, which will necessitate re-milling in order to get the material into the fine condition necessary for sowing.

The action of mixing is self-explanatory from the figure, and charging and discharging are effected by bringing the appropriate scoop into position. Mixing is effected in half a minute. The drum is rotated by means of a pair of toothed rings, one at each end of the drum, which gear with a pair of small pinions rotated by means of a pulley. It will be noted that the material is not subjected to any rubbing, grinding or squeezing action liable to cause the material to paste and harden.

The machine is made in batch sizes of 5 cwts., 10 cwts. and 1 ton, corresponding to mixing capacities of 13, 27 and 60 cubic feet respectively. The output varies in the different sizes from 3 to 20 tons per hour, and absorbs from 3 to 10 brake horse power. Speed of revolution is 120 r.p.m. in the smallest size, 100 r.p.m. in the intermediate and 75 r.p.m. in the largest.

This mixer is excellent for preparing relatively small batches of accurate composition for the fulfilment of small orders. Owing, however, to the necessity of ageing mixed fertilisers for periods varying from a week up to six months, it is necessary to anticipate large orders and make for stock. Fertilisers which give the most trouble with caking are those containing sulphate of ammonia and superphosphate. If bagged immediately after mixing the whole sets to a hard, rock-like mass perfectly useless to the farmer. This tendency is greater the greater the content of sulphate of ammonia up to a point. The usual practice is, therefore, to mix sufficient of a high-grade fertiliser of this character containing the maximum quantity of sulphate of ammonia likely to be required. The heap is allowed to mature for as long a period as possible, but certainly not less than a month and preferably two months. From this heap,

after maturing, other mixtures containing
less sulphate of ammonia can readily be
made and bagged at once, ready for delivery.

The mixing of these heaps for stock is
usually effected continuously in a type of

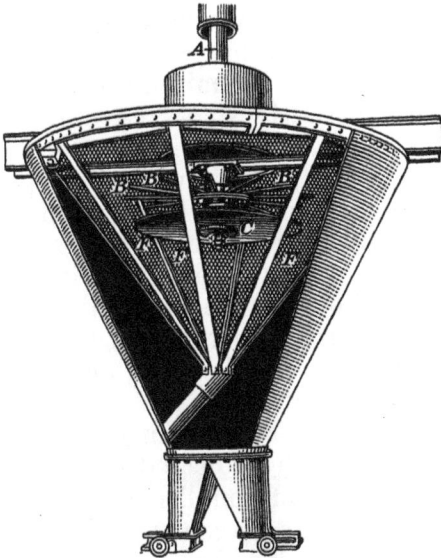

FIG. 16.—PULVER BLENDER.

machine which pulverises by a series of
rapid blows upon the unsupported material.
One example of such a machine is Tyler's
Pulver blender made by the Bickle Engineer-
ing Co., of Plymouth, and illustrated in
Fig. 16. The vertical, balanced shaft, A,
carries beaters, B, of a particular shape.
Below the beaters is a disc, C, which also

rotates with the shaft. The material is fed by means of an elevator (not shown in the diagram) to a chute, whence it falls over the distributing cone, which gives even distribution of the material to the beaters. The shaft revolves at a speed suitable to the hardness of the material, giving a linear velocity at the tips of the beaters of 6000 to 8000 feet per minute, whereby the lumps are instantly shattered and hurled against the screens, F, which separate the fine material. Lumps which escape the beaters are caught upon the disc and hurled outwards upon the screens with sufficient force to disintegrate them. Finally fan blades (not shown in the diagram) are fitted below the plates and rotate with it, thus circulating the air and drying and aërating the pulverised material.

Owing to the screens being inclined, the area of each mesh projected on a vertical plane is considerably smaller than the actual area and therefore screens of a large mesh which will not readily choke up, can be used without passing too coarse a product.

The machine is, of course, both a pulveriser and a mixer. When used as the latter the materials are brought to the foot of the elevator in the correct proportions and simultaneously shovelled into the boot, to be elevated, pulverised and blended as above described. The tailings run down to the bottom of the screens and are returned by means of a chute to the boot of the elevator.

These machines are made in small portable sizes with capacities from 1 to 4 tons per hour, and in large fixed sizes with capacities from 10 to 20 tons per hour. Power required in the small sizes is from 5 to 10 h.p. and in the large 15 h.p., including that for the elevator. These machines may be used for the preliminary mixing preparatory to ageing in the heap, and for this purpose are highly efficient, as the aerating and drying effect produced by the fan blades upon the hot, damp superphosphate is very beneficial in getting the material into fine, friable condition. For the purpose of mixing orders for dispatch, the small portable sizes are especially useful, as they may be moved about from one part of the store to another so as to deal with different heaps.

Another type of machine consists of a pair of shafts carrying rows of arms; the arms upon one shaft pass between, and thus partially mesh with, those upon the other. The shafts are revolved in opposite directions at high speed and the material is fed between them and thereby pulverised and blended.

In all these machines it will be noted that the main principle involves blows upon the unsupported material, and this, in fact, is the only satisfactory means of dealing with mixed fertilisers in lumps. Any attempt to deal with such materials otherwise generally proves unsatisfactory. An example of such an attempt may be cited to show its futility.

In order to blend sulphate of ammonia with superphosphate continuously, the former was delivered by a table measuring device (see Chapter III) on to a tray conveyor which carried it along in front of the mouth of the den from which the superphosphate was being excavated mechanically. Thus the latter formed a layer superimposed upon the ribbon of sulphate of ammonia. The two materials were delivered over the end of the conveyor into the boot of an elevator, whence they were carried up to the top of the building and delivered, by means of a hopper, on to a steel belt conveyor. A diagonal, laminated steel blade mounted upon a carriage whereby it could be moved to different positions along the belt was provided, to deflect the materials off the belt so as to drop them at any point on the ground below its line of travel. It might be thought that after so much travelling, handling and falling the materials would be well mixed by the time they reached the deflector blade; this was not, however, the case. To begin with, they could easily be distinguished as separate pockets of material in the buckets of the elevator, being hardly mixed at all; upon the belt they were partially mixed, but not nearly sufficiently. Finally, the deflector blade actually separated them to a great extent, by rolling the superphosphate into balls about the size of a pea, and these balls rolled down the side of the heap as it was built up and collected at the bottom, and the result was a heap in which the superphosphate was mainly at the bottom, and the finer, crystalline, sulphate of ammonia at the top. The result, as a mixing, was perfectly useless and had to be milled two or three times before it was properly mixed, in a type of machine previously described, in order to break up the small hard balls of superphosphate and thoroughly blend them into a homogeneous mass with the sulphate of ammonia.

This case indicates the necessity of proper treatment of fertiliser materials in order to mix them thoroughly, and the importance of avoiding as far as possible in the process any action which tends to squeeze or rub the material.

In passing it may be mentioned that the recent tremendous explosion at the works of the Badische Anilin- & Soda-Fabrik at Oppau in 1921, in which a large dump of a mixed fertiliser consisting of sulphate and nitrate of ammonia exploded, causing appalling damage and loss of life, was attributed at the inquiry to the existence of pure ammonium nitrate which detonated the whole mass during the process of breaking up by blasting. The existence of such pockets would indicate imperfect mixing, and the explosion would probably not have occurred if this had been properly carried out, as, although nitrate of ammonia can be detonated, repeated attempts to detonate the mixed salts in experiments at Woolwich failed.

The mixing of raw phosphate and basic slag in the preparation of " slag phosphate " or enriched slag has already been referred to in Chapter III.

CHAPTER VIII

Rubber Mixing. — The incorporation of various substances such as sulphur, lime, litharge, etc., with raw rubber for the purpose of vulcanising and filling, is an extremely important branch of the rubber industry. Sulphur is added in varying proportions according to the quality of vulcanised rubber required; lime and litharge are added as accelerators of vulcanisation; whilst various other substances, such as carbon black, lithophone, slate dust and bitumen, are added either as simple fillers to cheapen the product, or to improve its strength and texture.

Raw rubber is a stiff, plastic and somewhat sticky substance, and calls for special treatment in mixing. The ordinary type of mixer would be quite useless, as the rubber would simply cling to the blades and shaft as a sticky mass and revolve with them, no mixing action whatever taking place. The materials, or "drugs," as they are termed in the trade, might be incorporated by dissolving the rubber in some solvent such as naphtha or carbon disulphide and then adding the "drugs." In this connection, the colloid mill may be applied and extremely intimate mixing obtained. The cost of evaporation and recovery of the solvent would, however, prohibit the use of this method for large and heavy goods such as tyres. The means universally employed for this purpose consist in squeezing the materials into the rubber by means of a pair of cast-iron rolls, set close together; the distance apart may be varied. These rolls resemble those used for washing the raw rubber, but are plain, instead of being fluted or corrugated. They are made of cast iron, and are hollow, to enable them to be heated or cooled. They revolve in opposite directions at relative speeds varying from 3 to 2, down to 1 tooth difference in the pinions actuating them. The usual size is 36 inches long by 16 inches diameter.

The process of mixing is as follows. The rubber is passed through the rolls repeatedly, the latter being gradually tightened up during the process, until it is thoroughly " masticated " or plasticised and formed into thin sheet. The materials, in fine powder, are then sprinkled upon the sheets, which are again passed through the rolls whereby the powders are pressed into the soft mass. This process is repeated until a homogeneous plastic mass, in which the " drugs " are evenly disseminated, is produced. Any rubber which adheres to the rolls is removed by a scraper or " doctor." A batch of 1 cwt. takes 30 to 40 minutes to mix.

The aim is to produce a homogeneous mass or " dough " without overworking or " tiring " the rubber. Here again, as in

the mixing of the superphosphate in fertilisers, we have a material of peculiar
character which calls for special treatment.

The rolls are usually direct driven by
electric motor, but occasionally belt drive
is employed from a line shaft.[1]

Hydration of Lime.—The mixing of lime
with water to produce hydrate of lime is a
process which calls for special treatment
where a satisfactory product is required.
The process may be divided into two classes,
according to whether a wet hydrate (*i.e.*,
milk of lime) or a dry hydrate is required.
In the former case no special features are
presented; a circular vat provided with a
rotating agitator somewhat resembling a
blunger or wash-mill is the apparatus
required. This is filled with water and the
quicklime added in lumps. The lumps
quickly take up water and hydrate, with
production of heat and swelling, and consequent disintegration. The arms of the
agitator quickly disseminate the fine hydrate
first formed on the surface of the lumps
throughout the liquid, exposing the cores
to the further action of water until completely hydrated. The milk is run through
a screen to a storage tank in order to separate stones and any pieces of unhydrated
lime. In some cases the lime is placed on a
grid or in a basket kept in vibration, which

[1] See "India-Rubber and its Manufacture," by
H. L. Terry. Constable.

is immersed in the tank. The movement of the basket separates the hydrate from the unhydrated lime and impurities, whilst the agitator serves to hold it in suspension.

Where a dry hydrate is required, a different problem is presented. Water must be added in only just sufficient quantity to form the hydrate. Too much will produce a paste, whilst too little will give some quicklime in the product, which, for most purposes for which the hydrate is required, would be highly undesirable and deleterious. The mixture passes through a sticky or plastic stage before being converted to powder; consequently a type of mixer combining the functions of both wet and dry mixing is required.

Hydrators may be of the batch or continuous type. A simple type consists merely of a horizontal cylinder perforated with holes. The quicklime is added at one end, and as it travels through the cylinder, meets jets of water. Hydration takes place, heat is evolved and the dry hydrate is screened out through the holes, whilst impurities remain behind and are worked out at the opposite end, the cylinder being slightly inclined for this purpose. The disadvantage of this type lies in the difficulty of making the holes the correct size. If too large, unslaked lime and pebbles will find their way into the product, and if too small they will quickly become clogged.

The " Clyde " hydrator consists of a circular pan of similar type to that of a pan mill, but in place of rolls, scrapers are provided which plough and turn the material over, breaking up and exposing fresh surfaces to the action of the water until hydration is complete, when the batch can be removed and screened. The pan is surrounded by a hood and stack to carry off steam and dust. Another type, the Weber, consists of a trough containing two horizontal shafts carrying paddles or beaters revolving in opposite directions, with a discharge door at the bottom. This machine resembles Fawcett's Tar Macadam mixer (see Chapter V).

Both the former are of the batch-mixing type. Turning to continuous hydrators, of which the rotary screen previously mentioned is a type, two may be mentioned.

The " Kritzer " hydrator consists of a series of six tubes placed one above the other. Each tube contains a spiral conveyor of the open or interrupted worm type. The lime is fed in to the top tube, whilst water is sprayed in at a point two-thirds of the way up a stack which surmounts the whole system; in this way a portion of the fine dust is caught and returned to the system, some steam condensed and the water warmed; it flows down into the tubes and hydrates the lime, and the hydrate falls out at the bottom.

The "Schaffer" hydrator works on a similar principle, but is of different design. It consists of an upright cylindrical vessel divided into circular chambers by means of shelves. A central, vertical shaft carries arms with rabbles which sweep the shelves. The latter has holes alternately at the centres and peripheries. The lime is fed in at the top and is swept down from shelf to shelf, the water being sprayed into the stack. The whole closely resembles a Herreschoff mechanical furnace.[1]

A type of mixer recently introduced for a variety of special purposes, and adaptable to the above purpose, is the "Fusion" mixer made by the Fusion Corporation of Middlewich. It consists of a horizontal cylinder in which lies an elongated paddle having four or six blades. This paddle, which extends nearly the full length of the cylinder, merely lies therein and has no other support. The action of this paddle is shown in Fig. 17. In *A* it is being raised by the rotary movement of the cylinder to the position shown *B* from which it rolls over into the position *C*. By this action the material is continually turned over and over, and subject to a beating and slicing action, whereby it becomes thoroughly mixed, whether wet or dry, and prevented from adhering to the sides and caking. The

[1] See "Lime and Magnesia," by N. V. S. Knibbs, Messrs. Ernest Benn, Ltd.

blades do not come into direct contact with the sides, but fall upon the bed of material being treated. Movement of the material through the cylinder is effected by the slight " head " between the feed and discharge end, and not by sloping the tube to even a slight extent; if this were done the paddle would tend to work down to the lower end.

Mixing in the Organic Chemical Industry.— The manufacture of organic compounds

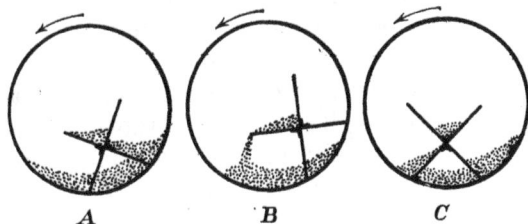

FIG. 17.—CROSS SECTION OF FUSION MIXER.

involves many processes such as nitration, sulphonation, and acetylation, in which mixing problems arise. Here the object of mixing lies in bringing the various materials into more or less intimate contact in order to promote and facilitate chemical action, and the type of mixing apparatus chosen will depend upon the nature of the materials and the type of chemical action involved.

Processes of nitration and sulphonation will involve the mixing of liquids with other liquids either soluble or insoluble, or with

solids. In the former case, when the liquids
are mutually soluble a very simple type
of mixer will suffice; such is the case in
the sulphonation of phenol. A deep, cylindrical vessel jacketed with steam has suspended in it a vertical shaft carrying at the
bottom an agitator in the form of an inverted,
hollow frustum of a cone, which is attached
to the shaft by means of webs or vanes
arranged spirally.[1] On rotating the shaft
this agitator draws the liquid from the
lower layers and circulates it, thus effecting
mixing. In the case above mentioned the
sulphuric acid is placed in the sulphonator
first and the agitator set in motion. The
phenol, previously melted, is then added,
and dissolves almost at once with the production of the orthosulphonic acid. If
the para-, or the di-sulphonic acid is required
the whole must be kept in gentle agitation
at a temperature of 100–110° C. for several
hours. . As the phenol is already in solution,
and consequently in molecular contact with
the acid, the agitation actually plays no
part in bringing the materials in contact, but
merely serves to cause the whole volume of
liquid to pass in close proximity to the
heated walls of the vessel, and thereby to
become more rapidly heated throughout.

Where the liquid to be treated is insoluble
in the acid, such as in the sulphonation of
benzene and its homologues, more vigorous

[1] See Chapter II.

agitation is required in order to disseminate
the benzene throughout the acid in as fine
globules as possible. The benzene only
dissolves in the acid just as fast as it reacts
with the same, and consequently agitation
must be maintained until solution is complete. For this purpose, an agitator is

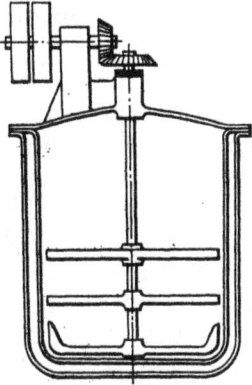

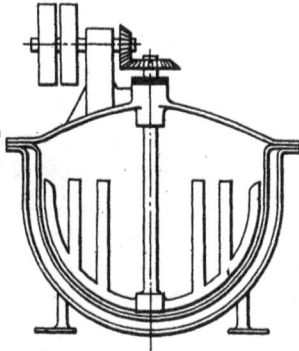

FIG. 18.—HORIZONTAL   FIG. 19.—VERTICAL ARM
ARM AGITATOR.   AGITATOR.

required with arms extending nearly to the
full diameter of the vessel, such as is illustrated in Fig. 18, which shows a mixer with
horizontal arms. In the case of a round-
bottomed or hemispherical type of mixer,
an agitator with vertical arms such as is
illustrated in Fig. 19 is more convenient.
This is a highly efficient type, as the semi-
circular blade sweeps the bottom of the
vessel and prevents deposition. Mixers of

this pattern are, therefore, highly suitable for the sulphonation or nitration of solid materials.

A very useful form of sulphonator stirrer consists of a vertical shaft carrying two impellers one above the other; the upper one drives the liquid downward and the lower one upwards, whilst arms are mounted between the two to increase the agitation.

In many cases it is desirable to have a tilting vessel in order to discharge the contents. In order to provide this with a vertical shaft agitator which can remain in position when the vessel is tilted, the arrangement shown in Fig. 20 is convenient. The vessel tilts upon the hollow trunnions, AA, which are concentric with the sprocket B, which drives bevel gearing, C, by means of the chain, D, and thence the peculiarly shaped agitator, E. The whole of the agitating gear, including the bevel gear, horizontal shaft, F, and sprocket G, tilts with the pan, the driving chain moving round the small sprocket B, and the agitator may be kept moving as the vessel is tilted.

The carrying out of reductions such as of nitrobenzene to aniline, or in fact any reaction in which a metal such as iron, in the form of filings or borings, or zinc dust enters, necessitates a highly efficient agitator which will sweep the bottom of the vessel and maintain the metal in a constant state of suspension and prevent it settling.

In the laboratory, such a reaction is usually
carried out using a much greater proportion
of hydrochloric or acetic acid than is used
on the works, consequently a much greater
proportion of ferrous salt is present to
effect reduction, and vigorous agitation is
not necessary.   The small quantity of acid

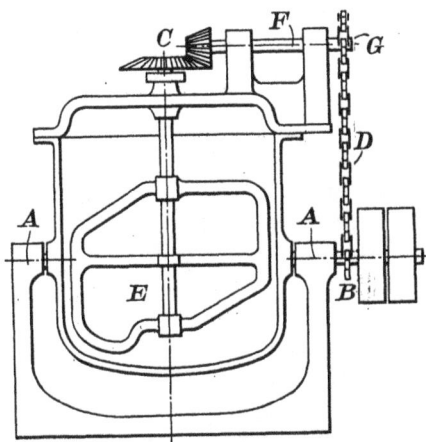

FIG. 20.—TILTING PAN WITH VERTICAL GATE AGITATOR.

usually employed in works reactions of this
kind, however, involves a much smaller
concentration of ferrous salt.   This is constantly being oxidised to the ferric state, and
in turn being again reduced by the iron
present to the ferrous condition, and the
rate of reduction will depend upon the
number of times the iron can be oxidised
again in a given time.   This in turn depends

upon the efficiency of the agitation. Some
of the agitators previously described are
suitable for this purpose, but an example of
a highly efficient type is the Planet agitator
illustrated in Fig. 21. The upper bevels,
A, drive the vertical shaft, B, which in turn

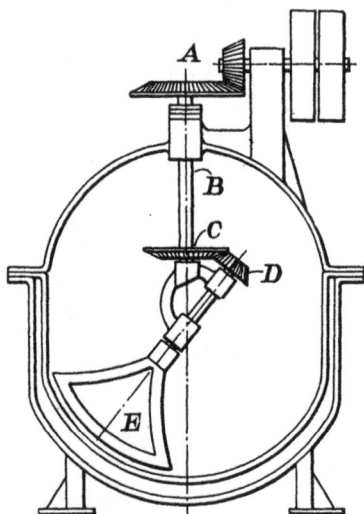

FIG. 21.—JACKETED PAN WITH PLANET AGITATOR.

drives the lower crown wheel, C, and rotates
the agitator as a whole; at the same time
the rotation of the lower crown wheel
actuates the small bevel, D, and agitator,
E, which rotates and at the same time
travels round the vessel, thus effecting a
highly vigorous and efficient mixing and
agitating action.

Practically all these machines are provided with jackets for the circulation of steam, water or oil for heating and cooling, or with coils for the same purpose; in some cases both are provided, the jacket for heating and the coil for cooling. They may be lined with a homogeneous lead lining, and the agitators may be similarly covered with lead. Messrs. Paul Schütze & Co., of Oggersheim, specialise in this class of work, and also in steel coils cast into the body of the vessel in place of jackets, and in steel mixing vessels lined with cast iron.

Mixing in the Soap Industry.—The soap industry employs a variety of mixing machines termed crutchers, from the fact that mixing in the early days of the industry was carried out with wooden poles or crutches. They are used for blending the various materials with the stiff soap. Such materials include fillers such as detergent clay, perfumes, colours, etc. A very common type consists of a vertical, cylindrical, jacketed vessel containing a vertical shaft agitator; the latter comprises a hollow cylinder concentric with the shaft and joined thereto by a spiral web. The action of this web is to force the soap upwards or downwards through the cylinder according to the direction of rotation, and return it between the cylinder and the walls of the vessel. Drive is effected through crown and bevel wheels; of the

5

latter, two are provided which mesh with the crown wheel, and either may be locked to the shaft by means of a sliding dog clutch so as to revolve the agitator in either direction; by this means extremely intimate blending is effected. Sometimes a pipe coil surrounds the agitator for heating purposes.

For thin soaps the agitator is provided with arms to prevent the outer layers remaining more or less " stagnant." [1]

Mixing in the Paint and Varnish Industry. —A few types of paint mixers have already been described in Chapter III. The principles of paint manufacture involve a preliminary grinding of the pigment in oil in a pan mill, followed by fine grinding between triple rolls, or in a cone mill; both these processes effect mixing; sometimes an intermediate process of mixing is effected in a simple type of horizontal mixer, designed to keep the whole contents in motion so that a homogeneous cream is fed to the finishing rolls.

Mention may advantageously be made of the Torrance edge runner or pan mills made with positive action to the rolls. In the usual type of pan mill the rolls revolve by simple friction between them and the pan, or rather the material in it; this is suitable for rough or dry materials which set up considerable frictional resistance; but

[1] See *Modern Soap and Detergent Industry*, by Geoffrey Martin.

in the case of pigments in oil, when the whole has been ground down to a certain consistency there is a tendency for the rolls to skid round the pan without revolving, with the result that mixing is not properly effected. To avoid this, the Torrance mill is fitted with a positive drive from pan to roll; the latter carries a · toothed wheel having teeth projecting horizontally from the rim; these teeth mesh with another pinion fixed to the pan and revolving with it in a horizontal plane so that the roll is positively driven by the pan and is independent of friction with it. This positive gearing is designed with such ratios as to revolve the roll or rolls at a greater speed than they would if revolving by simple friction with the pan. The materials are thus subjected to a shearing as well as a crushing action, and the effect is the same as would be produced by two or three rolls running at different speeds; in addition, the twisting action of the rolls, continually drawing the material from a straight into a circular path, ensures an extremely thorough grinding and mixing action. Spring ploughs turn over and deflect the material into the path of the rolls, further increasing the mixing action.

These mills are made in varying sizes from $2\frac{1}{2}$ to 8 feet in diameter. They are under-driven by crown and bevel pinions. Discharge is effected automatically through a door operated by a handle opening into a chute.

The positive gearing between pan and rolls is protected by means of a ridge on the pan surrounding the pinion, and in addition the whole gear may be enclosed.

The inside of the pan and the exterior of the rolls are ground perfectly true, and polished so that the smallest particles are subjected to their action when they come in contact. Moreover, it is not necessary to true up the surface from time to time, as the wear is even, the surfaces continually trueing one another up.

The function of the finishing rolls is to grind and mix the paint to an extremely smooth and fine cream. They consist of three rolls of granite or roughened steel, revolving at different speeds in opposite directions. The paint is fed on to the slowest roll, is carried round and taken up by the intermediate roll which revolves at a faster speed, and thence on to the third roll, which is the fastest, from whence it is taken by means of a scraper or " doctor," as it is termed. In addition to a rotary movement, the central roll oscillates from side to side in order to even the wear.

Messrs. Torrance and Sons make an improved type of mill of this character. The rolls are driven by gears machine-cut from the solid and running in an oil-bath. Adjustment is effected by a micrometer down to a fraction of a thousandth of an inch between the rollers; by this means

any desired pressure may be put upon the
material without actually bringing the
rollers in contact, and thus preventing any
passing; by this means also a single passage
through the rolls will reduce the paint to
the desired consistency.

By a special arrangement of gearing,

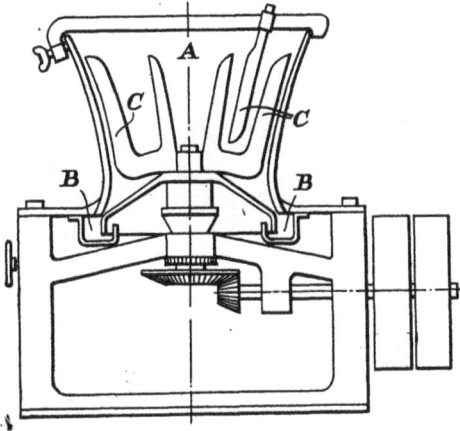

FIG. 22.—CONE PAINT MILL.
Adapted from Allen's *Chemical Engineering.*

alteration of the adjustment of the rolls
does not affect the depth of engagement
of the gear teeth, which always work at the
correct depth (*i.e.*, with their respective
pitch circumferences just in contact), thus
giving smooth and noiseless working with
minimum wear.

For the mixing of smaller batches of
paint, the cone paint mill illustrated in
Fig. 22 is suitable.

The conical bottom of the pan, A, is rotated by means of the bevel pinions

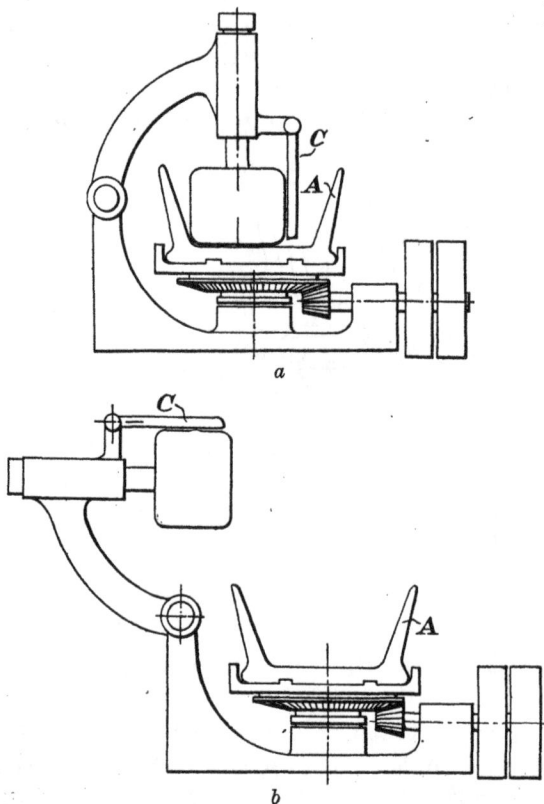

FIG. 23.—END RUNNER MILL.

mounted below, and the material escapes through the fine annular ring formed between the sides and the bottom into the trough, B; the latter is made with a sloping bottom so as to draw the paint off at one side. The width of the annular escape ring can be adjusted by means of a screw, not shown in the diagram, which raises or lowers the conical bottom. The mixing blades, C, revolve with the bottom and are detachable; they are used for working up stiff or thick materials.

Coming down to small machines for mixing very small batches for experimental purposes, the end runner mill, illustrated in Fig. 23, is very useful. The pan or mortar, A, which may be made of Wedgwood earthenware similar to an ordinary laboratory mortar, is mounted in a housing from which it may be quickly detached for cleaning. This housing is revolved either by hand or power through the bevel pinions. In (a) the pestle is shown in the working position, and rotates by friction with the pan; the material is thus subjected to a twisting action on the bottom which effectively grinds and mixes it to a fine paste or cream. In (b) the pestle is shown swung out of action; the scraper, C, removes any material which may adhere to it.

The Torrance mixers described in Chapter III are also suitable for mixing small or medium-sized batches, and also for thinning down thick materials.

INDEX